DREAMS IN BLUE

Railways and Social Change on the Clyde

COLIN TURBETT

Whittles

First published in 2026 by Whittles Publishing, an imprint of Porto Press.

British Library Cataloguing in Publication Data
A CIP record for this book is available from the British Library

ISBN: 978-1-84995-733-5

Cover and text design by Raspberry Creative Type, Edinburgh

Printed and bound in Great Britain by CPI Group (UK) Ltd, Croydon CR0 4YY

To order please go to our website www.portopress.com or contact our distributor, BookSource, 50 Cambuslang Road, Clydesmill Industrial Estate, Glasgow G32 8NB. Telephone 0141 642 9192

Porto Press Ltd
3 Connaught Road
St Albans
AL3 5RX

www.portopress.com

Paper from responsible sources

CONTENTS

FOREWORD

Glasgow is one of the biggest cities in Europe. It is one of its grandest, most hardworking and culturally vibrant. It is a fascinating, intricate and hardy place full of heart. Its wider urban region encloses a population of nearly two million people. This is not a city that should be limited by the lack of ambition that has hampered almost every other British conurbation. Indeed, Glaswegians today are kept moving by the UK's best suburban railway network outside of London.

Yet the city is at a junction. Choices are being made now as to the ceiling of ambition of the next step in Glasgow's mobility story, and there is a serious risk that – not for the first time – the spirit of this city will be contained rather than unleashed. Before us, the scale of the Clyde Metro remains undefined.

What better a time than now, then, for us to step back and consider the path taken so far? There can be no clearer guide for the opportunities that imagination can create.

As a prominent advocate for the empowerment of our cities and for the ambition of our politicians to lead rather than trail the ambition – and need – of citizens, I'm delighted to have been given the honour of opening Colin's detailed and determined account of Glasgow's dance with modern mobility, represented as any viable integrated transport system should be by high-density electric railways.

It is a deep frustration of mine that electrification is seen by so many in the UK as a subject to be argued for or against, when in mainland Europe an unelectrified railway is generally considered incomplete, and the idea of a diesel suburban railway is laughably quaint. Yet the trials and tribulations of a city on the front-step in that argument are instructive. Glasgow's success in the battle with central Governments in both Westminster and more recently Holyrood on the modernisation of its railway system forge a clear path for other cities to follow as a renewed zeal for urban devolution digs in for other parts of the UK.

But what of the bigger picture?

The Modernisation Plan, the British Transport Commission's expansive 1950s programme of railway renewal, enhancement and replacement, is often decried as a colossal, multi-billion-pound failure. This could not be further from the truth, and the prominent success of Glasgow's suburban electrification provides the perfect exemplar of this.

Indeed, anyone as cynical as I am might be led to believe that the long-sustained historical claim that the Modernisation Plan was a failure is a very useful canard to undermine the enormous success of the Plan's programme of electrification that sent electric tendrils across the country from the late 1950s right up until the arrival of the wires on the West Coast Main Line into Glasgow in 1974.

No. Electrification quite literally accelerated the success of the railways wherever it was deployed, not because of the benefits of electric traction for the railway, but because of the immediate improvements in the quality of service that electric trains brought passengers. Clean, quiet and quick.

And in Glasgow, unlike the rest of the country where the expansion of urban electrification and all that it brought stalled, the momentum largely kept up, with electric services expanding through the late 70s into the early 90s. If the privatised era brought stagnation, Scottish devolution returned somewhat of a keenness to the raising of wires above tracks, and since the mid-2000s the route mileage of electrified railway in Glasgow has nearly doubled.

In laying out the narrative of Glasgow's suburban railways, from their development as a bypass for poorer tenement developments through to their deployment as a lifeline for all of Glasgow's population, Colin lays out before us a distilled version of the wider story of Britain's cities. Post-war optimism paved the way for a fracture in approach, with individualised mobility on one side and a densifying of public transport on the other. Glasgow saw both in extremely heightened form: wide motorways tore through the urban fabric as the Blue Trains began snaking out into more of the Glaswegian hinterlands. In parallel with each other, the Glasgow Inner Ring Road and its connecting motorways grew, and the electrified suburban railway network grew alongside it.

This, then, is not just a story of rail. It is a story of the competition between individual and collective mobility, and it is a story in which we see how only collective mobility can truly endure. Glasgow's city centre is slowly pushing the motor car to its fringes. Serious consideration is being made to the abandonment of the M8 north of the Clyde, that barely half a century ago was carved through the north-west of the city inflicting so much gruesome misery on its urban fabric.

At a time of contracting employment as Glasgow underwent aggressive and unchecked de-industrialisation, the investment in the city's future through the 1950s right up until the early 1990s now seems quite exceptional. Again, this demonstrates that amidst plenty of political wrangling, there were those with the vision and the capacity to deliver change who handed the city and its hinterland an asset that it could not live without today. That spirit must now guide the development of the Clyde Metro.

Meanwhile, the design legacy of the first Glasgow Suburban Electrification, honoured to this day by a series of plaques still visible in Queen Street station, potentially steers the look of the entire British railway network today.

In the aftermath of the popularity of the Blue Trains, it wasn't just their paint colour that was emulated (with a touch more black and green in the mix) to create British Rail blue, but it is quite possible that the short-lived overlapping arrows identity of the 1960 electric services inspired the iconic British Rail double arrow.

It is the double arrow – now in its modern form as Rail Symbol 2 – that endures as the strongest symbol of Britain's future reintegrated railway system. But this symbol without an energised (literally and figuratively) railway cannot alone rescue Britain's aspirations for collective mobility.

The 1955 Modernisation Plan was responsible for a transformative boot up the backside of a largely decrepit system, and electrification was at its heart. In 2026, as Great British Railways rises in whatever form it finally takes, electrification must be at its heart once more.

As I write this, the world is beginning to comprehend the consequences of the biggest energy shock in 50 years, and possibly in human history. The scale of the crisis is incomprehensible to most of us. Those parts of the world best placed to survive it are those places that have ensured that its denizens can move around without relying on expensive personal transit.

The last energy shock close to this scale, also driven by decisions made around Iranian sovereignty, resulted in a renewed emphasis on railway electrification and revived a post-Modernisation programme of electrification that put wires between Edinburgh and London on the East Coast Main Line and was only stifled by the privatisation of the railways in the early 1990s. The railway must once again rise to the challenge.

Whether it is electrification on a national scale or the expansion and densification of suburban services on a city-wide scale, Glasgow's late-20th-century ambition gives us a blueprint not only for how the railways will fix the future, but how they can solve the problems we face today, right now.

Take it away, Colin.

Gareth Dennis, October 2026

ACKNOWLEDGEMENTS

The author has received considerable assistance with the research necessary for this book. Those who have given of their time and knowledge include (in no particular order): Darrel Hendrie (ScotRail driver and leading light of the Glasgow Electric Group who have preserved a Blue Train), Paul Strathdee (retired BR Scottish Region Traffic Controller), Jimmy Law (retired BR staff), John Binning (SPT Policy Officer who kindly allowed me to view the small collection of past documents held by them), Dugald Cameron (artist), Stuart Mackay, Angus Doyle, Keith Richmond (ASLEF), Colin Browning (WW2 army education PhD student). Thanks also to the helpful staff of the Glasgow Mitchell Library, National Library of Scotland and the York National Railway Museum Archive.

Photos and images are credited wherever possible and licences for use obtained when required. These are all copyright unless indicated as found in the Creative Commons domain. If any images have been inadvertently used without the appropriate permission, the author should be contacted so that the matter might be properly rectified.

Darrel Hendrie driving a 303 Blue Train on the Balloch Branch in their final months of service 2002 (Darrel Hendrie)

ABBREVIATIONS

ABCA – Army Bureau of Current Affairs
AC – Alternating Current
APT – Advanced Passenger Train
ASLEF – The Associated Society of Locomotive Enginemen and Firemen
ASRS – Amalgamated Society of Railway Servants
ATP – Automatic Train Protection
BEMU – Battery Electric Multiple Unit
BR – British Railways (pre-1965) / British Rail (post-1965)
BR (Sc) – British Railways/Rail Scottish Region rebranded as Scotrail (later ScotRail) in 1983
BRB – British Railways Board
BREL – British Railways Engineering Ltd
BTC – British Transport Commission
BWP – British Way & Purpose
COP – Conference of the Parties
CR – Caledonian Railway
DC – Direct Current
DMU – Diesel Multiple Unit
DOO – Driver Only Operated
ECML – East Coast Main Line
EGIP – Edinburgh Glasgow Improvement Programme
EMU – Electric Multiple Unit
GAAP – Glasgow Airport Access Project
GARL – Glasgow Airport Rail Link
GCT – Glasgow Corporation Transport
GDP – Gross Domestic Product
GGPTE – Greater Glasgow Passenger Transport Executive
GGTS – Greater Glasgow Transportation Study
GSWR – Glasgow & South Western Railway
GWR – Great Western Railway
HST – High Speed Train
KW – Kilowatt
LMS – London Midland & Scottish Railway
LNER – London & North Eastern Railway
LRT – Light Rail Transit
NBR – North British Railway
NSHEB – North of Scotland Hydro-Electric Board
NUR – National Union of Railwaymen
RCA – Railway Clerks Association
RCH – Railway Clearing House
RMT – Rail Maritime and Transport Workers
SNP – Scottish National Party
SPAD – Signal Passed at Danger
SPT – Strathclyde Partnership for Transport
SPTA – Strathclyde Passenger Transport Authority
SPTE – Strathclyde Passenger Transport Executive
SVR – Severn Valley Railway
TOC – Train Operating Company
TSSA – Transport Salaried Staffs Association
TUCC – Transport Users' Consultative Committee
UK – United Kingdom
V – Volt
VPF – Value per Prevented Fatality
WCML – West Coast Main Line

INTRODUCTION

Blue Train newspaper advert 1961
(Darrel Hendrie Collection)

This is a book about railways, and 'Blue Trains' in particular, but it is also about social change and post-war reconstruction in Glasgow – an area of Britain synonymous with heavy industry, endemic poverty and overcrowding – but through its socialist traditions, hope for a better world. It's a book about aspirations and their realisation through an infrastructure project that aimed to change lives and point to a brighter, cleaner future. That might sound familiar to the reader who hears today's politicians make promises about such ideas and projects as 'HS2', the 'Northern Powerhouse', the aim of Net Zero carbon emissions, and the need to overcome the idea that the UK is full of areas far from the London centre that get overlooked and forgotten.

As a boy who became fascinated by railways in the 1960s, I was struck with the design and deployment of the Blue Trains – they were spacious and comfortable, and passengers were offered fine views from the seating. All this was in strong contrast to the usual experience at that time, including electric trains elsewhere. It was no surprise that the blue livery later adopted by British Rail followed (but in more subdued form) the Caledonian Railway colour that gave the Blue Trains their name. Their appeal was summed up for me by the British Railways' poster (a painting by the artist Terence Cuneo) that graces the cover of this book. With its grimy steam engine on one side and a busy River Clyde at Bowling on the other, it gives period context to the Blue Trains' introduction.

The story starts in the Second World War with the early realisation by Government and social planners that a population being asked to make huge sacrifices, including that of their very lives, to fight fascist aggression, needed more than talk of King and Empire if morale and fighting spirit were to be maintained. What the armed services and those backing them at home needed was a vision of a better country – one worth fighting for. With this in mind, and with the backing of politicians of all persuasions, social planners and propagandists got to work: the result was a series of policy commitments and ideas exchanges that would lead to a Britain where, in the words of Sir Ernest Beveridge, the five giants of want, disease, ignorance, squalor and idleness were to be things of the past[1]. Whilst the foundations of the welfare state were being laid down, others were working on projects that would create work and prosperity. This was a different vision of Britain, one where the state would intervene to provide security and share the fruits of prosperity, hitherto enjoyed only by the wealthy and privileged, to ordinary people across the country.

Tom Johnston, the old Clydeside socialist and now one of the architects of this future, chaired a commission looking at using the natural resources of Scotland, mountains and rain, to create hydro-electricity that would bring work to the Highlands and power to the masses wherever they lived[2]. It wasn't quite Lenin's formula that communism would be created from Soviet power plus electricity, but it was revolutionary in its own way. The conservatism and suspicion of state intervention that had clouded social policy for generations was to be replaced by a new Britain where the needs of the population would come first. Whilst Churchill provided the gutsy war leadership the country seemed to require when its back was against the wall, victory in Europe in 1945 brought all these ideas and aspirations to the surface: a Labour Government was elected on a landslide, with a strong mandate to create a National Health Service and a welfare state, reinvigorated public education for children, and sweeping changes to the infrastructure of the country through its programme of nationalisation.

This change naturally included the railways. By the end of the war the system was worn out and people were tired of their experiences of overcrowded, slow and late running trains. Whilst some elements of modernisation had come about in suburban areas of English cities in the 1930s (particularly in the south-east and with express services to the capital), for most of the population, the network seemed Victorian and dated. In Glasgow, as in most cities and towns throughout the UK, there was an extensive electric tram system developed in Victorian times which came (perhaps mistakenly) to be seen as anachronistic in the post-war period due to a growing obsession with road transport and cars. There was also a massive slum clearance programme to be considered in the city which led to large sections of the population being moved out to new towns and peripheral housing schemes – with attendant public transport issues although the railway network was still extensive. In keeping with the promises of the post-war Labour Government to bring about popular reform, the railways were nationalised on 1 January 1948, ending competition and, it was hoped, duplication and waste. The railways, like the mines, docks and road transport, now belonged to the people – at least in theory.

The housing and transport infrastructure changes that began to take place in the 1950s and 1960s took place against a series of official planning initiatives at local, regional and national levels that were not always quite in sync with each other. The outcome of urban planning in the Glasgow region was generally welcomed – many of the new living spaces were less intensely populated, often sited near green areas, and included such luxuries as inside toilets and baths. Problems emerged that beset these areas to the present day, many due to the overlooking of obvious aspects of social planning (like shops, pubs and other social gathering places) that were evident long before the houses and new communities were planned and built.

Meanwhile, and back to electricity, power was mainly generated by coal-fired power stations – Johnston's hydro power had its limitations and would never meet the nation's needs in total. Coal was dirty, dangerous to extract, and its use lent a pallor to the atmosphere that contributed to smog and health problems. A suburban railway journey in Glasgow until the 1950s and 1960s was a grimy affair with windows tightly shut – especially through the long tunnels that featured on the lines to and through the City Centre. The need to eventually replace coal's extensive use on the railways was recognised by the four railway companies pre-war, and modernisation plans in the newly nationalised network after 1948 included electrification of local, regional and inter-city lines on what became a greatly reduced network. Electricity was, of course, mainly generated by coal, but by the time this form of motive power became widely employed in the latter quarter of the 20th century, power generation was increasingly achieved through other more sustainable, or at least less unhealthy means.

A recent academic paper has extended this premise, of electricity replacing coal, to argue convincingly that as the railways developed around steam-driven

Train entering Anderston Cross, Glasgow Central Low Level Line, July 1959
(© Transport Treasury: WAC Smith)

motive power, they were also used to transport coal to every corner of the country to fulfil domestic energy needs in most households – in effect an national energy grid system.[3] 1963 saw the completion of the National Grid that carried electricity across Britain. Now the requirement for the railways to carry coal to thousands of locations where it was taken to people's homes by local coal merchants, was no longer such a necessity. The Beeching rationalisation of that era, leading to widespread closures was not apparently based on that change, but was no great coincidence. The age in which a huge volume of railway freight traffic was devoted to the very slow transport of coal was over, justifying the drastic reduction in the network but freeing up the remaining lines for much faster services. This marked the dawn of British Rail, Inter-City, and electrification of the main lines, but the end of the railway system as we had known it. In the current era electrification is valued as a contribution to Net Zero and the Carbon-Free transportation ambitions of successive governments. The Blue Trains, in every respect, were ahead of their time.

Scotland can apparently claim the earliest experimentation with electric rail traction: according to an article in *Railway Magazine* of August 1931, an experiment with electric power was conducted on a section of the Edinburgh to Glasgow line in November 1842. This involved a carriage powered by electromagnets which achieved a speed of 4 mph over a one-and-a-half-mile stretch. The inventor, Davidson, claimed that a larger battery would produce better results, but his ideas seem to have progressed no further at that time; his call for better battery technology echoes down to present times.

The Glasgow suburban electrification whose passenger services began in 1960, had its origins in the 1955 British Railways Modernisation Plan as well as wider post-war development proposals. Constructed over a few short years using engineering methods and manual labour that seem primitive and almost criminal in their poor attention to health and safety, they were nonetheless forward looking, and the railway community were rightly proud of their labour and effort to bring this revolution about. The initial introduction of the Blue Trains was short-lived due to poorly tested engineering innovation, but when sorted out and services resumed, they proved popular with an increase in passenger use and timetable availability. By the end of that decade, electrification had spread to other areas in the Glasgow and Clyde area, and getting on for 60 years later, is still being extended. Although steam is long gone, the replacement diesel traction on the non-electrified lines proved expensive in its reliance on fossil fuel, and is now considered obsolete and unclean just like its coal predecessor.

This is not a technical book (although some description cannot be avoided), nor is it of appeal only to those seriously interested in railways – it is rather a social history surrounding a particular technical development aimed at the general good (that said, those train-buffs who love Blue Trains will find Chapter 5 of particular interest). The following chapters will outline the developments explained briefly in this introduction. This is a people's history ending with the present day but pointing to the future. Primacy though, goes to the Blue Trains – an arguable realisation of the wartime dream of a better future and a symbol of modernity and sound urban planning to meet the needs of ordinary people. The Blue Trains of 1960 stand out in the author's view as a historical milestone and achievement in terms of infrastructure development that we should marvel at today.

Like the contemporary railway engineer and author Gareth Dennis[4], I believe that railways have the potential to make this ailing world a better place, reducing inequalities and opening up horizons for mobility and travel that are not socially and environmentally destructive. This is far from their origins in the industrial revolution and the replacement of waterpower by coal-derived steam power. The development of the railways through the course

Old and new at Partick Hill 1961 (© ColourRail)

of the 19th century was piecemeal and driven by competition and greed for profit. It was based on the exploitation of resources: that of coal and that of human labour – both negative in their application as it turned out. Coal and other fossil fuel extraction has resulted in disastrous climate change that threatens our future, whilst the labour practices of the 19th century led to a world of inequality that persists into this century – and gets worse. Railway staff in the UK have achieved some improvement in relative terms through the strength of their trade unions, but we have a long way to go in making our rail system a significant part of our sustainable future, partly because we are saddled with haphazardly built infrastructure from another age. Railway engineers and planners continue to grapple with that despite the uncertainties and insecurities of successive funding arrangements and this is evident in Glasgow, where electrification marches on over 80 years after it was first suggested during the Second World War.

I'm sure that it was common for people of my generation, as we grew to adulthood, to take for granted the relative peace and world order built by our parents' generation in the wake of their wartime experiences. Perhaps we now realise how precarious that security and improvement in social conditions was. The progress characterised in 1960 by the Blue Trains and the pride taken in them by passengers and railway staff in the West of Scotland demonstrates what can be done when various factors come together – when the stars align.

It is in the spirit of celebrating positive aspects of our recent past and seeking a better future that this book is offered to the reader.

Chapter 1

HISTORY OF RAILWAY DEVELOPMENT IN GLASGOW

Introduction

Having already stated that this is not a railway book in the traditional sense, its central place deserves some background in railway history in Scotland. The railway network in Britain in the 19th century grew to meet the requirements of the industrial revolution: moving goods and people, especially, around the expanding cities where wealth and manufactured items were being created from transported raw materials. Later routes were developed based around leisure and escape from urban squalor – in Glasgow alongside and complementing steamer shipping to the coastal resorts of the Clyde. The piecemeal and competition-driven fashion of railway development was to create problems for future transport planners – cheaper adaptation of existing infrastructure usually winning out over fresh and futuristic planned development based on existing and expected need. This chapter will look briefly at how this all took place in the Glasgow conurbation and how competition was replaced by public ownership that began to point the way to planned development.

Glasgow's Early Railways

Glasgow's growth had started in the previous century through its trade based on slavery and raw material created elsewhere – sugar, tobacco and textiles. In the 19th century its proximity to coal and iron resources led to its expansion as a centre of engineering and manufacturing, including railway locomotives, that were the subject of profitable overseas export. In this respect, as a major seaport, shipbuilding became of great significance.

Glasgow's population grew from 77,000 in 1801 to 760,000 a hundred years later, and reached a peak of 1,100,000 in 1925. The city sucked in those displaced by economic change in Ireland and the Highlands of Scotland, creating intensely populated areas in the city where this vast pool of workers lived and worked. Overpopulation, lack of sanitation and rampant disease resulted in some of the worst urban conditions found in Europe, with half of children born dying before the age of five. This led to urban expansion as the better off moved out to the West End and Southside of the city, but tenement building continued apace, leading to the city becoming one of the most densely populated in the world by the turn of the century. Improvement in conditions had come with the supply of fresh water from Loch Katrine and investment in sewage disposal, and the city continued to grow with the absorption of burghs on its borders. The need for railways to move people and goods was now greater than ever, accelerating their development as profitable sources for investment. [1]

The first mainline railway was the Garnkirk and Glasgow Railway opened in 1831, followed that decade by lines that spread from the city to Edinburgh, Kilmarnock, Greenock and Ayr. In the 1840s Carlisle, Aberdeen and Dundee were reached by rail. In 1848 a through service was introduced between Glasgow and London, followed two years later by a slightly faster direct route to London from Edinburgh via the east coast. By 1864 the total mileage in Scotland amounted to 1,720 miles, but by 1912 had stretched to 3,627, accounting for an annual 89 million passengers and 67 million tons of goods, bringing in an income of £18 million. The fierce competition surrounding this rapid growth took its toll, and amalgamations were a regular feature as the bigger fish swallowed the smaller ones.

Between 1850 and 1866 three companies emerged who would dominate the Glasgow area and central Scotland for the next 60 years: the Caledonian Railway Company and the North British Railway whose lines extended across the country, and the smaller Glasgow and South Western Railway (GSWR) whose territory was more localised giving them a near monopoly in the south-west of Scotland. The three railway companies were all in competition and their developments and services overlapped. But running powers over each other's networks did facilitate goods and passenger transfers. There were also smaller joint railways in the Glasgow area operating as independent companies over which the three ran trains. Competition was fiercest on the lucrative seasonal Clyde coast routes resulting in races to get people to sea and lochside destinations by rail and steamer. Although vessels left from the

Broomielaw in Glasgow, such was the stench from pollution in the River Clyde within the city that people preferred to go far downstream by rail to join them in towns nearer the cleaner open sea.[2] Tramway systems were being developed by the end of the 19th century, winning custom from what was now a well-developed rail system, with the result that closures of stations in the Glasgow suburban rail network started well before the First World War and continued apace long before the rationalisations and closures of the post-Second World War and Beeching eras.

Railway People in the 19th and early 20th Centuries

As mentioned importantly at the close of this book's introduction, the rise of the railways took place against two forms of interlinked and negative exploitation – the replacement of waterpower to drive rotational machinery by the efficiency and flexibility of the coal-fired steam engine, and the exploitation of human labour at the lowest possible cost to the investor at all stages of that process. Coal and fossil fuel extraction, as we have always known but have only recently come to acknowledge, contributes to rapid and potentially catastrophic climate change.[3] Whilst many of us retain a nostalgia for the steam train, few of us would want to associate that with gross exploitation of labour, from children being employed in coal mines to long working hours and the hazards of railway employment in the early years. The two though are indelibly linked with the historical development of the industrial revolution.

A fact of interest when researching for this book was that whilst information relating to engineering and infrastructural aspects of railway development was easy to access, the human stories rarely went beyond the characters of leading engineers, directors and capital investors.[4] Statistics quoted frequently concerned passenger numbers and freight tonnage, and workforce detail seemed of little interest despite the railways being major steady employers in every corner of the country. As with so much written about the Victorian period and history up until fairly modern times, the workforce who created the wealth of the nation through the industrial revolution is often marginalised and overlooked – giving way to heroes, almost all men who are celebrated in marble in our cities and parks, as well as the written word. The scale of railway employment and organisation can be seen in this snippet: on 7 September 1899, the Caledonian Railway organised an outing to Carlisle from their St. Rollox engineering works in Springburn,

Glasgow. This huge works built steam locomotives, carriages and wagons for the company. The outing transported some 15,000 workers and their friends in 14 trains that left Glasgow at regular intervals starting at 05:10, all running to time including on the evening return journeys.[5] The Caledonian Railway employed 29,316 staff (all grades) in 1921, the North British 31,869 and the Glasgow and South Western 12,142, contributing to a United Kingdom total (excluding the companies in the island of Ireland) of 735,870.[5]

It would be wrong to characterise the Victorian era of railway development as one of paternalistic employers reigning over an acquiescent workforce. In Scotland these were years of great struggle in all three of the companies described in this chapter. The railway mania period of the 1840s saw unsustainable levels of speculatory investment and a bubble that inevitably burst so that fortunes were lost and the larger companies swallowed up the failed ones. The survivors naturally tried to maintain profit by squeezing the workforce, and this was resisted from early times. The Locomotive Engine Drivers' and Firemen's Amalgamated Benefit Society, which evolved into the Associated Society of Locomotive Enginemen and Firemen (ASLEF) was formed in 1867; the Amalgamated Society of Railway Servants (ASRS) was formed in 1872, and through various amalgamations via the National Union of Railwaymen (NUR), eventually became the Rail Maritime and Transport (RMT) union of the present day. Both suffered ups and downs as they sought to improve conditions and pay. During a wages' strike of Caledonian Railway employees in 1873, members report long hours to earn a basic wage – '...90 hours in five days', 'last week ... one day 20 hours and another day 24 hours'. Accidents were inevitable and frequent: in one month (September 1873) the Caledonian reported 36, and the previous year railway companies reported 1,145 deaths and 3,000 injuries – half of which involved employees. The usual outcome was denial of compensation as the victim was blamed if they were in the workforce.[6] The situation had only marginally improved by 1920 when the UK's total for staff fatalities and injuries came to 376 and 4,414 respectively. The most dangerous occupation was that of maintaining the permanent way – accounting for 105 of the deaths and 205 of the injuries; shunters were also at high risk of death and injury, as were firemen on the locomotives.[7]

Mention should also be made of the trade union for administrative and clerical staff: the National Association of General Railway Clerks was formed in 1897 and soon changed its name to the Railway Clerks Association (RCA) and, with a broader membership, to the Transport Salaried Staff Association (its present name) in 1951. Although never a militant union in its early days, it managed to establish and organise despite fierce opposition from the employers.

Success for the unions was hard won and at great cost: the Scottish Railway strike of 1890 started on the North British but was joined by workers from all three companies, and met a determined response from all three to refuse all demands concerning pay and hours as 'the spirit of the men must be broken... the men must be made to surrender... the working classes must not be allowed to taste power.'[8] Although the Caledonian had a reputation as a relatively benign employer, their hard line stiffened resolve and probably prolonged the strike: in Motherwell, they announced the eviction of strikers and their families from company-owned housing, sparking a violent response by a crowd of 20,000, requiring the military as well as police to carry their plan through. The strike was eventually beaten back and its leaders (300 in the Caledonian, 200 in the North British and 30 in the Glasgow and South Western) were dismissed and 'blacklisted'. However, the unions were at last recognised, and in the case of the GSWR, a maximum working week of 60 hours was instituted.[9]

The ASRS called for nationalisation of the system as early as 1896, soon followed by ASLEF and the RCA. The ASRS were in the forefront of seeking political representation in parliament, first through sympathisers in the Liberal Party, and from 1900 the Labour Representative Committee which soon became the Labour Party. One of the first two Labour MPs was Richard Bell, the ASRS General Secretary. Railway staff were in the front line when it came to legislation around trade union activity: the Taff Vale judgement of 1901, when the ASRS was held financially liable for losses incurred during a strike, effectively outlawed industrial action, until it was reversed under pressure from a growing Labour representation in parliament in 1906. The ASRS were also instrumental in establishing the right of trade unions to have political funds to enable support for parliamentary candidates, won in 1912, just before they became the NUR. The landmark moment for the NUR came in 1919 when a nine-day national strike by half a million railway staff beat off attempts to reduce wages and reverse gains made during the war years. Despite concerted attempts by Government and employers to break the strike, a settlement favourable to the unions was reached.

This is all highlighted to demonstrate that railway staff have helped shape our nation as organised workers as well as through their efforts as employees.

The Caledonian Railway Company

The biographer of the Caledonian Railway, O.S. Nock, describes it as having 'character' with 'qualities of charm and colour and that romance that is invariably associated with the Western Highlands and with navigation on the Firth of Clyde'. Nock contrasted this favourably with the 'ruthless efficiency of its English based business partner, the London and North Western Railway'.[10] Later in this story we will hear how the Blue Trains of 1960 were painted out in Caledonian livery, no doubt an attempt to capture some nostalgia for that bygone age when railways were new and exciting.

The Caledonian was formed in 1845 with the intent of reaching England from Glasgow and linking with lines to the south. This objective was achieved in 1848, with lines linking Edinburgh, Glasgow and the London and North Western Railway in Carlisle, these joining at Carstairs. In its early days the system used three terminals in Glasgow. But, from 1879, developed Glasgow Central Station as its main terminus and hub, this being completed in the form we see today, in 1906. The Caledonian retained an important terminus at Buchanan Street serving lines to the east and north. Over the previous 50 years, an extensive network in central Scotland was complemented by lines to Oban, with a branch to Ballachulish, in the West Highlands, and lines in the Edinburgh, Perth and Angus areas connecting with the main line to Aberdeen. In the south-west, and in the competition already referred to with the Glasgow and South Western Railway, a line went out to the harbour and passenger terminal in Ardrossan. These were all developed to serve freight as well as passenger needs, but what characterised the Caledonian was caution and sound investment in infrastructure rather than a race to expand, resulting in the failures that beset other companies in that era; early financial problems in the late 1840s had seen lessons learned, and from then on shareholders received steady returns throughout the life of the company. The result of all this was iconic stations, such as Wemyss Bay, and locomotives, such as '123', that now rests in the Glasgow Transport Museum. All the work of innovative engineers and designers revered to this day.

Caledonian Railway Coat of Arms (Wikimedia Creative Commons)

The Caledonian operated a number of suburban lines in the Glasgow area; in 1906 those advertised were:

- Central Low Level to Balloch via Scotstoun
- Central High Level to Hamilton via East Kilbride
- Maryhill to Coatbridge via Low Level, Tollcross and Baillieston
- Buchanan Street to Hamilton via Stepps Road and Motherwell
- Rutherglen to Maryhill via Low Level, Kelvinbridge and Kirklee
- Central High Level to Central Low Level via Queens Park, Kirkhill, Tollcross and Glasgow Cross[11]

In addition, it ran regular services out to the steamer ports of Ardrossan, Gourock and Wemyss Bay on the Clyde Coast, to Dumbarton, and inland to destinations in Renfrewshire and Lanarkshire. In 1923, the Caledonian Railway Company became part of the London, Midland and Scottish Railway Company, but competition continued, now with the London and North Eastern Railway, both inheriting lines that often served the same places along almost parallel railway routes.

The Glasgow Central Railway

The principal Glasgow suburban line operated by the Caledonian that is of main interest for this book, the Glasgow Central Railway, received authorisation in 1888 and opened in stages from 1896. Taking as its example the successful and largely underground Metropolitan Railway in London, the company's aim was to relieve a full to capacity Glasgow Central terminus, and open up a north to south rail route that would enable goods to easily traverse the River Clyde. The initial proposal was for an elevated railway that would run over the city's streets as pioneered by the Liverpool Overhead Railway. However, the city centre business lobby soon forced the plans to change to an underground route.[12] Tunnelling involved extensive engineering to divert and overcome obstacles such as sewers, gas and water pipes, all at a construction cost of £200,000 per mile (over £23 million today).[13] The main Glasgow freight terminus at that time was south of the river at General Terminus Quay, so the new line would facilitate goods traffic northwards from across the border as well as facilitate coal and minerals from Lanarkshire and outposts in Ayrshire and the West Highlands in each direction. Access from the south to the Queen's Docks (opened 1874) on the north bank at Stobcross was also a key driver. However, this was more than just a means of crossing the Clyde; its extension through Glasgow's West End to what was then the northern suburbs opened up new suburban passenger possibilities. The Glasgow Central line ran from Bridgeton to Maryhill, with an onward freight branch to the large industrial area in Dawsholm where the Caledonian built a large engine shed. At Maryhill it joined with other existing lines and at its east end with existing lines from the west coast route at Rutherglen. It ran underneath the mainline terminus at Glasgow Central with a low-level station that was completely underground.

Glasgow Central Low Level Line construction (BR Sc *100 Years of Glasgow Central* author collection)

The line was seven miles long, five of which were underground. Despite heavy passenger use in its early years, the line was unpopular – being dank, dark and very smoky. The railway writer C. Hamilton-Ellis described this (and the North British line described below) as sulphurous 'fuming burrows'.[14] Apparently regular passengers used paper bags to open train doors and prevent soot from dirtying their hands. The subterranean stations in the city centre were poorly lit, but the Botanic Gardens station in the moneyed West End was stylish with two ornate Russian-style towers with clocks at its entrance on Great Western Road. Such was the overall criticism of the line that the Caledonian Railway proposed to electrify it as early as 1898, but the technology at this time was in its earliest stages (despite the extensive use of electric trams in Glasgow and elsewhere), and the freight usage would have required engine changes at each end, so the plans were set aside, and most passengers went over to the newly electrified trams instead. The line had closed completely by 1964 due to tram and then bus competition, but the section from Dalmarnock to Stobcross under the city centre was, as we shall see later in this book, reopened and electrified as the Argyle Line in 1979.

The North British Railway Company

Whilst the Caledonian Railway is associated with the West Coast Main Line (WCML) to London down the west coast of Britain from Glasgow, the North British Railway is linked with Edinburgh and the East Coast route (ECML). However, there was no clear east/west divide as the two rival companies had tentacles across Scotland and competed in both cities.

The North British was formed in 1844 with the object of linking Edinburgh with railways to the south at Berwick-upon-Tweed, a goal reached in 1846. The early aspiration of linking Edinburgh with a direct line to Carlisle and the north-west of England took longer, and what became known as the Waverley Route was fully opened in 1862. In 1862 the North British acquired the Edinburgh, Perth and Dundee Railway, and in 1865 absorbed the Edinburgh and Glasgow and Monkland Railways, giving them the fastest route between the two cities, and an extensive network of routes across central Scotland, throughout Fife and up to Perth. Financial problems arising from false accounting were major issues in the mid-1860s but were partially resolved in subsequent years by rapprochement with the Caledonian Railway, their main rival, but discussion about amalgamation in 1872 was ended by the North British and competition remained a feature in subsequent years. The main terminus for North British passengers in Glasgow was Queen Street Station (originally called Dundas Street), acquired with the Edinburgh and Glasgow Railway, and rebuilt with its curved roof (as seen today) in 1878. Innovations included one of the first uses of electric light in Glasgow. In 1894 one of the most ambitious railway schemes yet seen was opened – a direct route from Queen Street in Glasgow to Fort William in the West Highlands – 'Mountain Moor and Loch' as the lavish publicity of the North British put it – 100 miles of mostly desolate country.[15] A few years later this line, the West Highland Railway, was extended a further 40 miles to Mallaig, enabling the carrying of freshly caught fish speedily to the south to feed the growing taste for fish and chips.

North British Railway Coat of Arms (Wikimedia Creative Commons)

Although the North British focused its central belt attention on routes north of the

River Clyde from Stirlingshire through to Helensburgh in the west, it was also involved in the development of suburban routes in Glasgow. By 1906 those advertised were:

- Hyndland to Shettleston via Bellgrove
- Hyndland to Airdrie (South)
- Clydebank East to Milngavie via Queen Street Low Level
- Springburn to Maryhill Park
- Jordanhill to Hamilton via Shettleston, Coatbridge and Bothwell
- Hyndland to Hamilton via Mount Vernon and Uddingston
- Bridgeton Cross to Helensburgh via Singer
- Maryhill circle via Queen Street Low Level and Springburn
- Possil to Maryhill via Queen Street
- Bridgeton Cross to Kilsyth via Finnieston, Great Western Road and Maryhill
- Partickhill to Broomhouse[16]

Like its main rivals, the North British owned Clyde-based steamers that sailed to the same resorts from Craigendoran, just outside Helensburgh.

The City of Glasgow Union Railway and the Glasgow City and District Railway

The Glasgow-based routes listed above were achieved through two main initiatives. The City of Glasgow Union Railway involved a partnership between the North British and the Glasgow and South Western Railway and bridged the Clyde enabling, principally, the easy transfer of goods between systems serving different areas of the country. Opened in 1874 but not reaching Springburn until 1887, the line ran from West Street on the joint main line south of the Clyde to Paisley, and Sighthill on the Edinburgh to Glasgow line, bridging the Clyde between Gorbals and Gallowgate, and facilitating through traffic. Six miles in length, its construction involved the demolition of 1,300 houses.

The Glasgow City and District Railway was similarly short but expensive and ambitious, this time designed to enable trains to run from west to east through the city centre, linking with the terminus of the North British at Queen Street, so running to the north of the route underneath Central Station. The line ran from Stobcross, adjacent to Queen's Dock and the lines westwards that ran down from the north of the city to High Street, which had been the College terminus. It was built as a 'cut and cover' line: through tunnels under the city centre with stations (Queen Street and Charing Cross) and was three miles in length; two and a half miles being underground. The line was opened in 1886, and like the Caledonian one just to the south was criticised for its dirt and smoke, to the extent that the cover of the station at Charing Cross was soon removed.[17] Although its beginnings were as an independent railway, the Glasgow City and District Railway was sponsored by the North British and to all intents and purposes part of its network. As we shall see later in this book, it became central to the 1960 suburban electrification project, and has been in continuous use since its inception.

The Glasgow and South Western Railway Company

The Glasgow and South Western Railway (GSWR) was formed in 1850 by the amalgamation of the two principal companies that had developed railways from Glasgow into the south-west of Scotland. GSWR trains into Glasgow initially terminated in Gorbals but after the construction of the City of Glasgow Union Railway and the bridging of the Clyde with a branch into St. Enoch in the city centre, the GSWR took over the terminus in 1883, making this their headquarters. Although Renfrewshire, Ayrshire and Dumfriesshire were always the focus of its enterprise, with resentment of incursions such as the Caledonian Railway route to Ardrossan, it went into partnership with the Midland Railway to develop passenger services from Scotland to England, the St. Enoch to London St. Pancras service surviving up until the 1960s. As with its rivals, the GSWR took advantage of the popularity of Clyde coastal steamer traffic – with routes linking with their own vessels, to Greenock, Largs, Ayr and Ardrossan.

GSWR suburban services in the Glasgow area were south of the Clyde serving Renfrew, Paisley, Barrhead and Kilmarnock on the route leading further out into the counties they served. The main line from St Enoch to Paisley Canal included stations in the south-west of the city but the domination of the other two

Glasgow & South Western Railway Coat of Arms (Wikimedia Creative Commons)

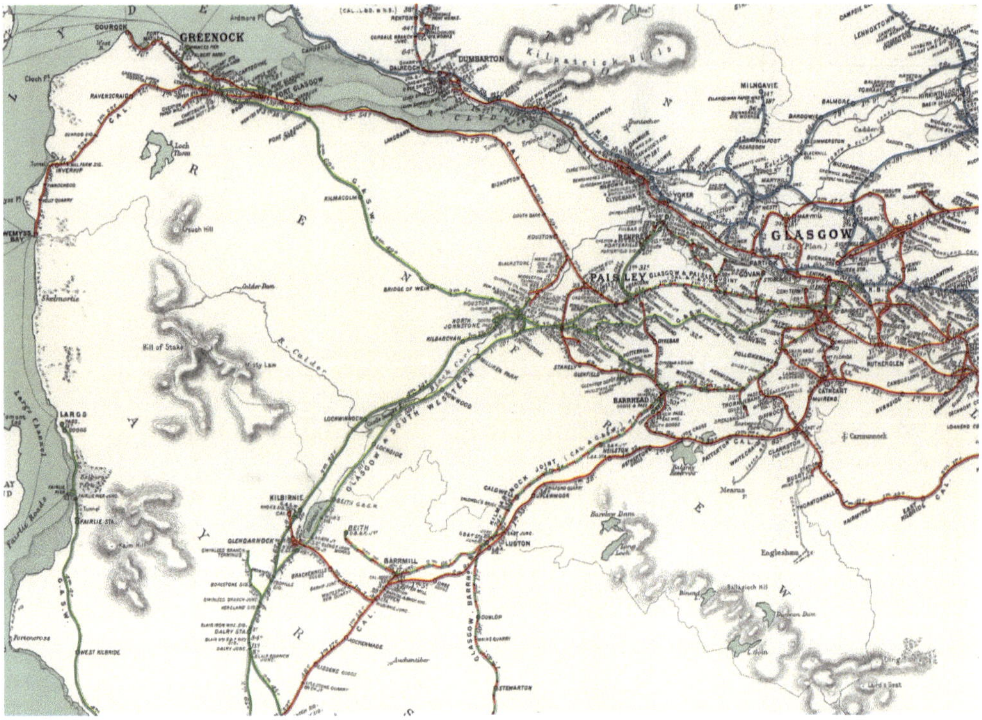

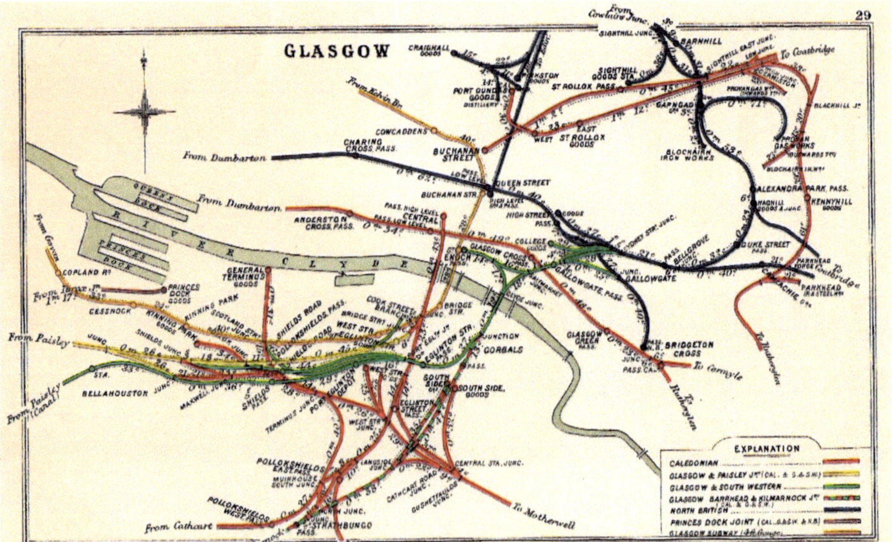

Top: Section of 1907 map showing the lines of the railway companies in the West of Scotland – Red = Caledonian, Blue = North British, Green = Glasgow & South Western (reproduced with the permission of the National Library of Scotland)

Bottom: RCH 1920 map of railways in Glasgow (author collection)

companies in the Glasgow area lent little opportunity to expand further. A branch line to Govan was an early victim of tram competition, closing several times before final withdrawal of services in 1921. The main route became more important with the development of housing in the south-west suburbs from the 1950s onwards but was not included in the electrification programme of the 1950s and 1960s.

The 1923 Railway Grouping

Financial difficulties in the wake of the First World War caused by wasteful competition and duplication resulted in a 1921 parliamentary act that grouped all the major companies in Britain (21 in total) into four super companies that were based on the major routes out of London – south, north-east, north-west and west. The railways were henceforth to be run in the public interest through the private companies. Government interference through such legislation would have seemed anathema prior to the war but had been necessary during the conflict to manage resources efficiently, and no longer seemed such an alien concept. Originally the suggested grouping had involved the establishment of a Scottish company, but this was resisted by the pre-grouping companies, by the unions who feared reduced pay and terms of employment, and passenger and freight interests who feared higher charges.[18] By this time the five companies in Scotland (the Highland Railway and Great North of Scotland Railway operating in the north of the country) were cooperating to a degree not generally found in England and Wales, but the changes were to be implemented across the whole country.

From 1 January 1923, the Caledonian (2,836 track miles) and Glasgow and South Western Railways (1,128 track miles) became part of the London Midland and Scottish Railway Company (LMS), and, with its prevalence in the east of Scotland, the North British (2,748 track miles) became part of the London and North Eastern Railway (LNER).

Top: LNER logo (Wikimedia Creative Commons)
Bottom: LMS Coat of Arms (Wikimedia Creative Commons)

The joint lines that had enjoyed an independent existence in the Glasgow area, being used by the Caledonian and GSWR in particular, also ceased to be. By this time the development of new lines was past: the railways were beginning to face competition from motor vehicles, and rationalisation that had begun with the closure of unprofitable lines began to gather pace. This spurred modernisation on the most profitable routes and efforts were made to increase speed, comfort for passengers and efficiency. The fact that all this might presage an attack on pay, terms and conditions for staff was not lost on the trade unions.

On the Southern Railway spreading out from its London hubs, an extensive programme brought about the world's largest network of suburban electrification (1,760 track miles) prior to the Second World War using a third rail direct current (DC) system.[19] Elsewhere, some suburban electrification had commenced prior to 1923, but in Scotland there were no such developments before the Second World War; this despite the recommendation of the Weir Committee of 1931 that the entire UK network should be electrified at an estimated cost of £261 million (almost £16 billion today), with completion over a period of 15 years.[20] Lord William Weir, the chair of the committee was a Clydeside industrialist, but even that level of influence made no lasting impact on the railway companies. Weir was far-sighted enough to realise that steam should not only be replaced but that its departure should not be marked by the growth of a dependency on imported oil.

A serious proposal was made for electrification of the LMS Glasgow Cathcart Circle, Neilston and Kirkhill lines in 1929. The scheme was convincingly argued for based on combating the growth of road and tram competition: the speed up that electrification and other improvement could realise would increase capacity by 200 per cent and bring back passengers. Extensive details of the scheme are given in Jack Kernahan's history of the Cathcart Circle: at a cost of £1.5 million these include the construction of 36 three-car train units, signal replacement, new workshop facilities and station facelifts, as well as 1500V DC electrification. Housing developments underway suggested that a 40 per cent increase in existing revenue would occur immediately and settle down to 10 per cent per year over the first ten years. The whole scheme could not be commercially justified as the world went into recession, and was passed over until after the Second World War.[21]

The express routes between London and Scotland by east and west coast main lines were in competition and both the LMS and the LNER introduced the fastest steam locomotives yet seen to reduce times on each route. Diesel locomotives, by now in widespread use in the USA due to the availability of cheap oil, were the subject of trial and experiment, but these and further electrification projects (the LNER line between Sheffield and Manchester via the Dearne Valley being the main example) where efficiencies dictated these as potentially profitable, were arrested by the outbreak of war in 1939.

Although the new companies were said to be keen to foster a corporate spirit amongst their staff by encouraging and sponsoring recreational and sporting activity[22], the economic and industrial climate of the 1920s was bound to have an effect on employees subject to a concerted effort to save on costs. The drivers' union, ASLEF, called out their drivers in January 1924 in opposition to wage reductions, but the NUR, under the moderate leadership of Jimmy Thomas MP, refused to join them, limiting their protest, like that of the RCA, to verbal condemnation of the employers.

On 3rd May 1926 the railway unions called out their members in line with a Trade Union Congress decision to support the locked-out miners nationally with a General Strike. The LNER did their best to break the strike and recruited over 4,000 'volunteers' in Scotland, using 734 of them as porters and locomotive staff. With normal safety rules and regulations

Volunteer strikebreakers, General Strike 1926 (Wikimedia Creative Commons)

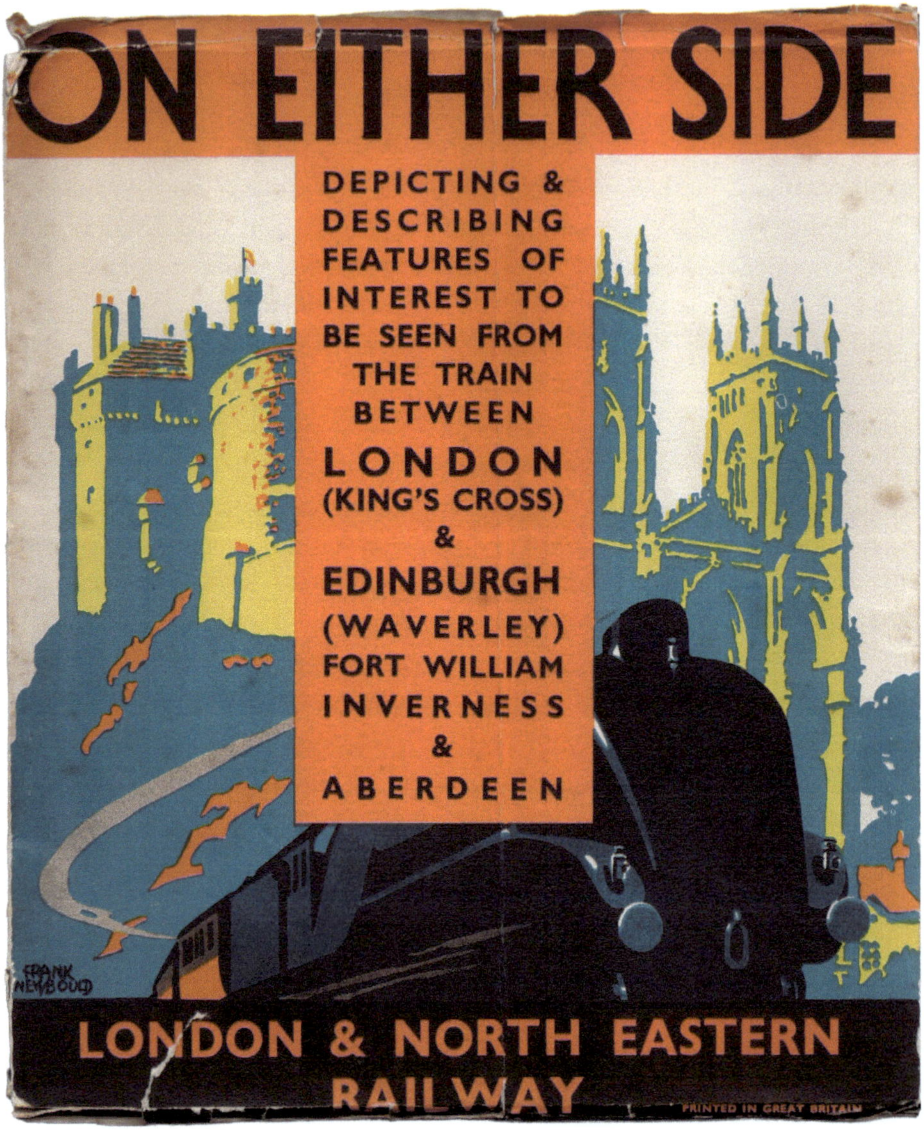

1930s Prestige Travel: Cover of the LNER guide to the East Coast Express Route from London to Edinburgh (author collection)

dropped for the duration, some of the strikebreakers were employed as drivers and signalmen – with inevitable results: a collision in Edinburgh resulted in three deaths. In areas where the strike was strongly supported, strikebreakers were stoned, including an incident in a Lanarkshire mining village when a crowd of 300–400 drove off a gang intent on track repair.[23]

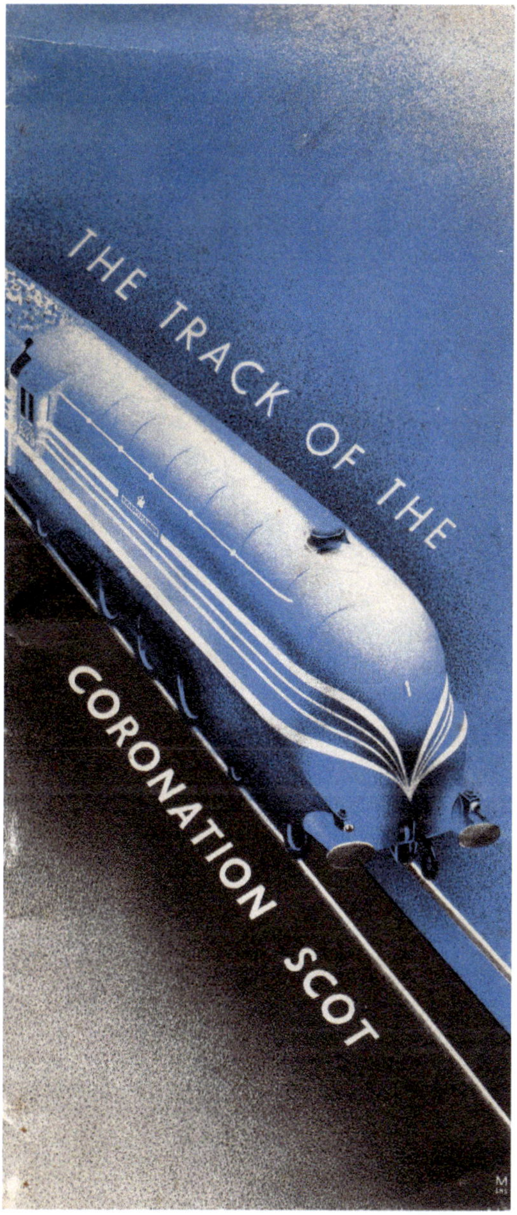

1930s Prestige Travel: Cover of the LMS guide to the West Coast Express Route from London to Glasgow (author collection)

The support of the NUR and RCA for the strike was half-hearted at best. Jimmy Thomas was a leading member of the General Council of the TUC and involved from the outset in negotiations with Government. Historians are generally agreed that he and his fellow council members were easily sold short by vague Government assurances to settle the grievances that had caused the strike. They called it off after nine days even though nothing was guaranteed. Until then it had remained solid. Walkden of the RCA went as far as advising his members that they had secured victory for the miners and that there would be no victimisations following return to work. This turned out to be very false: in October 1926 Thomas was forced to admit that across Britain 45,000 railway workers had not been re-employed three months after the strike finished, and that 200,000 were on a three-day week. For him this was proof that the strike was ill-conceived rather than, as others saw it, that his own actions had ensured defeat. The companies were blatant in using this as an opportunity to clear out militant trade union activists and reduce staff numbers.[24]

The 1930s saw a generalised economic crisis that affected every aspect of the economy and reached deep into people's lives. The duty to the Big Four shareholders meant that wage cuts and staff reductions followed upon income losses as demands for passenger and goods traffic fell. Competition with motor buses also increased substantially, leading to the demise of very uneconomic branch lines. Despite this climate there was some modernisation that fitted with increasing efficiency and saving on staff time: the underground line through Glasgow Queen St. Low Level was fitted with electric colour light signalling; St. Enoch Station main signal box became electronically operated; Glasgow Buchanan St. Station was rebuilt; and some single-track lines were doubled and automatic coaling equipment installed at some locomotive depots. Up until this point the railways in Britain had received no Government subsidies and with profits plunging whilst losses increased the railway companies lobbied Government for some equity, a 'Square Deal', given the amount being poured at that time into road building and improvement. Their request was for relaxation of the tariff system for merchandise transport put in place as part of the 1921 Railways Act (in principle dating back to the 1840s), initially designed to protect the interests of rail users and prevent monopolisation. The Transport Advisory Committee reported on this to parliament in May 1939, recommending that the railway companies charge as they see fit, but subject to the agreement of the Railway Rates Tribunal. Naturally the roads transport lobby wanted some say in all this, but the outbreak of war overtook expected legislation in 1940 and put the proposals on hold.

Tramway Systems – and the Glasgow Underground Railway

Glasgow once boasted the largest tram network in Europe and its development, like the railways, is linked with the industrial revolution and the Victorian boom in capitalism. However, it is also associated with late 19th-century municipalism and early public ownership, and the rise of the labour movement alongside growth of socialist traditions in the city.

Horse-drawn trams on standard gauge railway lines embedded in the city's streets and thoroughfares came to Glasgow in 1872, with the Glasgow Corporation owning the lines, but the investing promoters forming a company to lease the line and run the service for 23 years. Starting with a service from St Georges Cross through the City Centre to Eglinton Toll in the southside, within 20 years the Glasgow Tramway and Omnibus Company owned 2,000 horses, 233 tramcars and 24 buses, and were regularly developing new routes. The tramway workers were unionised from early on and in 1873 went on strike in opposition to their 15-hour day. Ten years later they were working

VIP's boarding the first electric trams at Springburn depot

First electric tram in Glasgow, Springburn 1898
(*100 Years of Electrification* SPT)

16 hours a day – described as 'petty tyranny and worse than Egyptian slavery' by the *Glasgow Weekly Mail* newspaper of 1 December 1883.[25] The late 1880s and early 1890s marked a period of industrial unrest throughout Britain (as we saw earlier with the railway strike of 1890) and the Glasgow Tramway workers reacted against their conditions and pay – now 20s a week for up to 17 hours work a day – a combination considered awful even by the standards of the day. The workers formed a committee of 20 to consider what to do and they were all promptly sacked. The recently formed Trades Council protested and called for the Corporation to take public control of the service, the lease of which would expire in 1895. The campaign was taken up by the tram workers' union representing carters and horsemen (what was eventually to become the Transport and General Workers' Union – and more recently UNITE the Union) and gained momentum as time went by. The nationwide Scottish strike by carters in 1889 lasted two weeks and resulted in concessions being won from some employers, and spurred the London Dock strike of the same year. In September 1889 the Trades Council organised a procession of 13,000 workers supporting the tram staff and their demand for public ownership. With trade union popularity and membership density growing, the campaign was ultimately successful, and Glasgow Corporation took over the system in 1894. The outgoing company refused to sell their equipment to the Corporation and tried unsuccessfully to run their own private bus service for a short period. Now under popular public ownership, the service took off quickly.[26]

Electric trams first appeared in 1898, requiring comprehensive infrastructure upgrading involving overhead wires, and the systems to support them, and more substantial street embedded rails for the heavier tramcars of which 400 new ones were constructed. After some experimentation with design, a four-wheeled double-deck standard tram was introduced – a thousand being built between 1898 and 1924. Electrification required a substantial power source, and a new coal-fired power station was built at Pinkston on the banks of the Forth and Clyde Canal, with substations around the system.

Pinkston was not linked to the National Grid until the 1950s, being owned until then by Glasgow Corporation for the sole purpose of supplying the transport system.

The upgrading required for the electrification also enabled some of the lines near industrial locations to be used by railway goods traffic, and this was the case in the shipyards, where Fairfields in Govan even employed an electric shunting locomotive from 1940 that drew power from the Corporation system.

Significant social progress with employment practices took place during the First World War: in 1915 a thousand tramway workers were recruited into the 15th Highland Light Infantry (Tramways) Battalion, resulting in the recruitment of women to take their place. By 1916, 1,180 women conductors were at work, and 25 drivers. Women 'clippies' as they were known, became an institution in Glasgow – on both buses and trams, their sharp wit and no-nonsense attitude famously characterising the city and its people.

The Glasgow Corporation tramway system extended far beyond the city boundaries: in the 1920s the Airdrie and Coatbridge Tramways, and the

Glasgow Bus & Tram network 1932 (*100 Years of Electrification* SPT)

Paisley District Tramway, were purchased by Glasgow Corporation, and services ran south into Rutherglen and out to Clydebank in the west. Altogether, at its zenith in the 1930s the system stretched to over 100 miles of tram routes.

Over the years tramcars were upgraded to give weather protection and the final generation, the Coronation Cars, were introduced in 1938, with a second version known as Cunarders in 1948. The decision to phase out trams will be discussed in a later chapter but, suffice to say here, this process began in 1949 and continued until the tram services ended in 1962 – the last such urban system in Britain until a revival in recent years in Edinburgh and some cities in England.

Glasgow's third form of railed transportation was (and still is) the Subway, known as the Glasgow District Subway until 1936 when it was renamed Glasgow Underground, and renamed again as Glasgow Subway in 2003. Constructed in 1896 by the private Glasgow Subway Company as an underground cable-drawn system, it ran for six miles in a circle under the Clyde connecting the west of the city on both sides of the river with the city centre, linking 15 stations. With no connection with any surface system, the Subway has always run on a four-feet gauge, making the carriages smaller than systems found in London and elsewhere in the world.

Like the Glasgow Trams, industrial action played a central role in the takeover by Glasgow Corporation and public ownership in 1923. In early

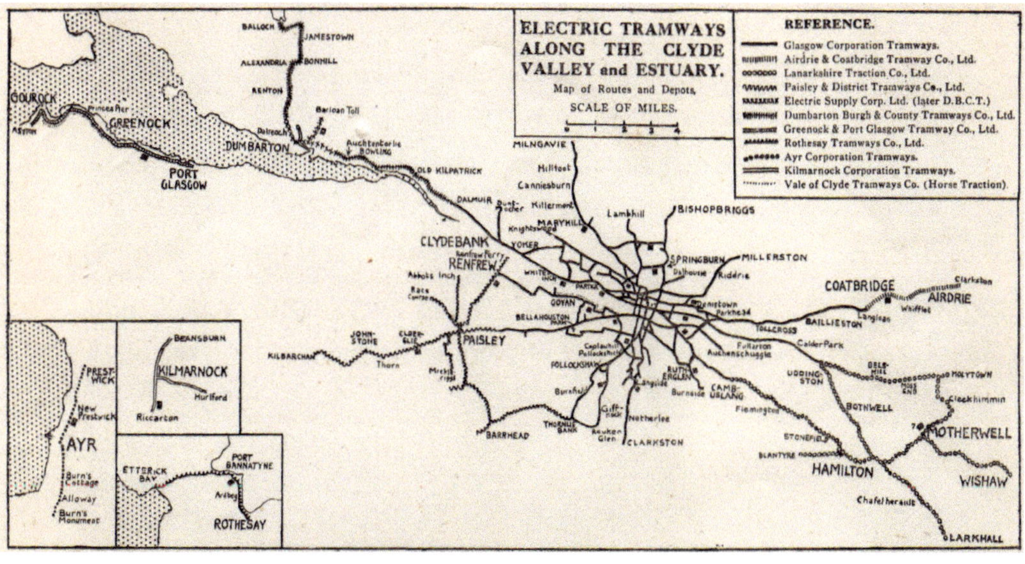

West of Scotland Tram Network in the early 20th century (RCH: author collection)

1920 the Workers' Union-led staff, seeking parity with tram workers, went on strike against a background of poor industrial relations and began picketing the workshops and system centre in Govan. They were backed and encouraged by the Govan socialist MP Neil McLean and some Corporation councillors. The strike was very bitter with picket-line arrests resulting in prison sentences for two men and a woman, of three months and one month respectively. After five weeks of strike action the company took on new employees and soon were back to a full service. However, the losses incurred could not be recovered in time to restore the confidence of the shareholders and the system closed completely in March 1922. The resultant chaos in public transport could not be addressed by increases in trams and cross-river ferry services, especially around Govan where thousands of shipyard workers had to get to and from their workplaces every day. The Corporation made an offer to the company that was accepted even though the shareholders only received 1/- in the pound and the official takeover took place from 1 August 1923 although services had resumed before this. The welcome given by the workforce was short-lived when the Corporation refused to re-employ 33 subway staff at the end of the ten-day General Strike in May 1926 – ten for picket-line activity, and the others for failure to pass a medical examination (i.e. reasons unconnected with the strike).[27] During the General Strike, as on the railways, some trams and buses had run with 'volunteer' strikebreakers, including Glasgow University students from privileged backgrounds who found such class warfare quite a jolly adventure.

The system was electrified with the addition of a third electricity-conducting rail in 1935, and has been the subject of upgrading and modernisation and rolling stock replacement in 1980 and 2024, but still runs through the original parallel tunnels on the original narrow gauge.

The Railways in the Second World War

Contingency plans for wartime operations had been in place long before the outbreak of the Second World War in 1939 and, as planned, on the first day of war under the Emergency Powers (Defence of the Realm) Act, the Government took over running the entire system, but leaving local management to the four companies – 'control but not possession'.[28] All revenue went to the Government, the controlling body that was created being the Railway Executive Committee, who paid the companies a fixed rate. In 1941, the Government created the Ministry of War Transport who provided more comprehensive central administration in the form of directions and instructions. However, total control was not considered to have been in place until the summer of 1944 and the huge operations surrounding *Operation Overlord* – the invasion of mainland Europe. From the first day of the war special trains were deployed to move mobilised troops and within a few days evacuate children in their thousands out of the cities. Demand was to increase as the wartime economy (armaments manufacture, coal production increases etc.), defence measures and invasion plans took hold during the next six years. By the end of the war, freight traffic had increased over the 1938 level, from 16,266 tons annually, to 22,023 tons; and passengers from 18,993 to 35,248. This was achieved with only a 4 per cent increase in locomotives, and 6 per cent increase in staffing, men in many instances being replaced by women.[29] In 1938 26,000 women had been employed on Britain's railways, most of them in work roles such as typists and in canteens, but this was now changing. Equal pay was demanded by the trade unions but rejected by the Railway Staff National Tribunal in 1942, the companies arguing that 'this was unsound on social grounds'[30] However, the following year equal pay was conceded; by this time women could be conscripted into railway work and 105,703 had been deployed in various roles – in fact almost every job on the railway apart from drivers and firemen whose work in those days involved a progression that took many years.

Crowded platform during wartime (Photographer unknown, *LMS at War*: author collection)

All this required huge commitment and effort from railway staff, who were subject to direct German attack as well as risk attached to the overburdening of the system and the strains placed on human operatives and equipment. Railway workshops were also turned over to war production, producing tanks, aircraft, bailey bridges, and smaller naval vessels as well as guns and a vast variety

Women cleaners at Camden Shed in wartime (painting by Norman Wilkinson *LMS at War*, author collection)

of components for fighting machines. The consequence of this was that maintenance of railway stock suffered and obsolete and worn-out equipment continued in daily use.

The impact upon passengers of wartime pressures was telling: for most of the war, Mass Observation's[31] National Panel determined transport issues as being one of the two major pre-occupations of the British people (the other being the blackout). Petrol rationing significantly reduced car and motorcycle usage, and compulsory deployment of workers across the country resulted in a 70 per cent increase on rail use. However, there were 10 per cent fewer carriages and tickets were three times the pre-war prices. There were few restaurant cars for longer journeys, no reservations and trains were often late if not cancelled altogether.[32]

With only basic maintenance to infrastructure and equipment (although turnaround was speeded up for wagon, carriage and locomotive repair), the systems nationwide were fairly worn out by war's end in 1945. In this, railway requirements joined a queue that included housing and production for peace to meet social and economic needs and live up to the aspirations of a victorious population who had survived the war. One factor in relation to railways was that the poor wartime experience of passengers had contributed towards a

Coal traffic was the mainstay of freight operations from the beginnings of the railway and was of vital strategic importance in the Second World War (MOI/HMSO *Transport Goes to War*, author collection)

lack of interest in the system; although the railways had moved mountains in terms of their contribution to victory, secrecy at the time meant that this was unknown to the majority of people. Their experience was of overcrowded, uncomfortable and slow trains, especially on the longer express routes – notably the WCML from London to Glasgow. This once prestigious route carried thousands of servicemen and women throughout the war under such conditions; these memories endured for a long time, and the line did not revive its former status until electrification was completed in 1974.[33]

The 1939–45 period saw great changes in attitudes and expectations amongst people at all levels of society, and one outcome of this was the 1948 nationalisation of the railways, a story that is the subject of the next chapter.

Chapter 2

WARTIME PLANNING FOR PEACE

> 'First of all, we say that we will get the kind of Britain in peace-time for which we are prepared to *unite* and *fight* to obtain. No one will give us anything. The new Britain will not be handed to us on a plate... It is true that the monopolists are already active in their demands for the ending of the vital controls after the war. On the other hand, we have seen, and the vision can never be wiped out, that in wither peace or war, a socialist economy is more effective and efficient than private enterprise can ever be.'
>
> FROM A STATEMENT BY UK TRADE UNIONISTS 1944[1]

Introduction

The Second World War saw a growth in determination amongst the British people that if the war was to have meaning for those being asked to sacrifice their lives it must be about defeating the threat of fascism *and* bringing about a new, safer and secure world in which the privations of pre-war life would be a thing of the past.

How the mood for change grew during the war years, and how that was manifested in future policy ideas are the subjects of this chapter. The worldwide war, only 20 years after an earlier European conflagration, involved sacrifice and loss on a scale that seems almost unimaginable today. Total war requirements accelerated scientific and technical progress for good and bad – who in 1939 would have imagined that a single weapon, the nuclear bomb, could unleash such death and devastation, shaping the world order for years after the war's end? Total war also involves the entire population: fighters, producers of munitions and survival requirements needed to secure victory, and of course potential victims now that death and destruction could be rained from the air anywhere.

Other authors have written of the contribution of the railways to victory[2]. Here though, is an examination of more general issues affecting how the railways might take shape in the peace to come.

Willing the Nation to Fight and Win

The ending of the First World War in 1918 did not deliver the land fit for heroes promised for the survivors. Disillusioned and traumatised by the pointless carnage of trench warfare, returning soldiers were met with the same class-based privilege and inequality that had preceded the conflict. The war years were followed by social tension and economic collapse as the system of capitalism and colonial exploitation, seen by many to have peaked before 1914, descended into chaos and destruction through the war and the unstable world that emerged in its wake. In Russia, Bolshevik revolution seemed to offer an alternative, and their vision of worker's power was attractive to many.

So too was emergent fascism with its vision of security through national pride, racist division and military strength. Trade union power in Britain, on the ascendancy in the period leading up to the First World War, was effectively broken by the defeat of the miners and their supporters in 1926, itself a reflection of the economic collapse that characterised the 1920s and 1930s. Unemployment reached startling levels in the 1930s, particularly in areas of 19th-century industrial growth, and was barely recovering when war broke out again in 1939.

The early period of war was relatively peaceful, and rearmament brought back some jobs, so there was some complacency until the shock defeat of the allied armies in the summer of 1940. Suddenly, Nazi invasion and military defeat seemed a possibility. The political crisis saw Churchill head off those who wanted to make peace with Hitler, and levels of uncertainty over the future of the nation reached new heights. The Dunkirk evacuation in June 1940 has since been treated in cinema and seen in the popular imagination as almost a victory. It was, though, in Churchill's own words, a 'colossal military disaster'.

The Ministry of Information (MOI) was established in September 1939, and immediately had to make up for lost ground if it was to counter the well-established propaganda machines of the totalitarian regimes with whom it was now at war. The methods adopted were borrowed from Hitler's *Mein Kampf*: appealing to the instincts of the masses rather than reason, using slogans and repetition.[4] The result was a set of defined objectives that governed the MOI's output of posters, pamphlets, films, broadcasts and press releases:

> To publicise and interpret Government policy in relation to the war, to help sustain public morale and to stimulate the war effort, and to maintain a steady flow of facts and opinions calculated to further the policy of the Government in the prosecution of the war.[5]

The general system of censorship ensured that agreed messages were propagated through the press (although some freedom of opinion was permitted), written media and broadcasting through the BBC. The latter became synonymous with facts and telling the truth about the war and was trusted as the main source of day-to-day information on the conflict's progress. By 1941, the MOI had learned that poster campaigns with exhortations to win the victory were not enough and gave them up. Instead, their output began to focus on news and information, designed to breed confidence and prevent harmful rumour. From then until the end of the war they churned out a succession of publications designed to popularise the war effort and celebrate the achievements of the fighting forces.

Cover of MOI/HMSO *Transport Goes to War, 1942* (author collection)

The MOI film unit contrasted the situation in the 1930s with the full employment that war had brought, pointing to the better and more productive society that would emerge from victory. This was echoed in carefully crafted mainstream films. The 1941 film, *Love on the Dole* portrayed a grim image of life before the war and ended with a message from the Minister of Labour, A.V. Alexander, that this must not be allowed to happen after the present conflict. The effect of celebrating the efforts of the armed forces and ordinary people on the Home Front through well-constructed words and images consolidated the feeling that this was a people's war, ultimately deserving of a people's peace. This began to overtake ill-feeling amongst those suffering from German air raids, mostly working-class people for whom the absence of adequate shelters and formalised aid was a cause of resentment at the height of the 1940/41 blitz. Morale was also held up as the war progressed through full employment with reasonable wages and the ready availability of food as a result of increased production and rationing.

Home Intelligence reports were made weekly to government from January 1940 – based on feedback from 13 regional centres whose task was to provide unbiased intelligence on what people were thinking and feeling about the war. These secret surveys of opinion demonstrated that many people were questioning what they were fighting for and in whose interests – whilst there was general support for ongoing resistance to the Nazis, the situation demanded absolute commitment to the war effort on the part of the whole population. Total war had no place for a half-hearted struggle as those not engaged in the armed forces would need to be involved in the production of war materials, food to feed the population and the services such as railways required to make all this possible.

Normal political debate was set aside when the main parties entered coalition under Churchill in May 1940, agreeing that a resumption of electoral politics would take place on war's end, but that unity was needed if victory was to be secured – a distant prospect at that time. This did not mean that debate ended, but the Government, through the MOI and BBC, made every effort to ensure that discussion was kept within parameters, and that priority was always given to the war effort. Fierce debate did emerge through the by-elections that took place, demonstrated by the election on successive occasions of 'Commonwealth' candidates pledged to post-war reconstruction and socialist politics, who beat coalition candidates in seats across the country from Tory shires to normally safe Labour seats.

Within the trade union movement there was a new confidence arising from the full employment brought about by the war economy. Although trade union leaders were moderate and had a significant representation in the wartime coalition, there were many shop floor activists determined to win lasting concessions for their members as they were urged to work hard with expected overtime to meet production targets. Shop floor activity had always involved militant socialists and communists and, after Russia entered the war in 1941, Communist Party members flipped from an anti-capitalist war position to one that made the people's war effort and support for Russia their main focus. They continued to propagandise, but for a communist future post-war rather than for immediate disruption and revolution. Their propaganda output urged workers to defer their struggle against the bosses until fascism had been defeated and work harder; a pamphlet aimed at Clydeside shipyard workers railed against those other socialists and anarchists who argued otherwise.[6] Strikes, which were illegal, did take place later in the war, notably in the Kent coalfield and amongst shipyard apprentices, but in general workers were earning well if working hard for a common goal in an atmosphere of national unity.

Left-wing socialist and communist opinion had long held that the railways should be nationalised and such an aspiration was now more in tune with the national mood, especially as (outlined in Chapter 1), they

LEFT: Trade Union pamphlet from 1944, looking to post-war (author collection)
RIGHT: Communist Party pamphlet 1943 (author collection)

were now Government controlled. The Communist Party, enjoying relative popularity thanks to its association with Russia who were bearing the brunt of the fight against Hitler's armies, produced a pamphlet in late 1943 arguing for better pay and conditions for hard-pressed railway staff, and also for public ownership and control, and the active involvement of railway workers in the running of the system. This was argued further in a 1944 pamphlet, *Transport for the People*, whose blueprint for nationalisation of transport was close to what happened in 1948. However, the Party called not for workers' control, but for trade union inclusion in 'Boards of Administration' that would operate at both national and local level.[7]

The call for nationalisation was not confined to the far left; the NUR did what it could to push the Labour Party into their eventual 1945 manifesto position of public ownership. They directly influenced Labour's 1943 pamphlet *Post War Organisation of British Transport* but within the union there were expressions of concern about the extent to which this might involve representatives of the workforce.[8] The pamphlet itself captured the mood of the times with its introductory words:

> 'Men and women in the Forces, away on war work or engaged in activities nearer home, are seriously concerned about the kind of world which will emerge from this war. They are not, in the main, concerned about theories, or for that matter with particular forms of industrial organisation, but how certain basic human needs and desires can be met...
>
> In considering the post-war organisation of transport, we must, therefore, have regard to the contribution which the industry can and should make to the development and social life of the nation.'[9]

The proposals from Labour, whilst perhaps vaguely worded and gradualist in respect of post-war nationalisation (so as not to upset wartime coalition partners), were clear about the need for co-ordination at all levels, and that passenger services should be integrated and in the hands of community representatives.

An influential book by a leading economist, Gilbert Walker, looked at the interface and competition between road and rail for freight carriage. It was published in 1942 but based on observations made prior to the war (a second edition was published on the eve of nationalisation in 1947).

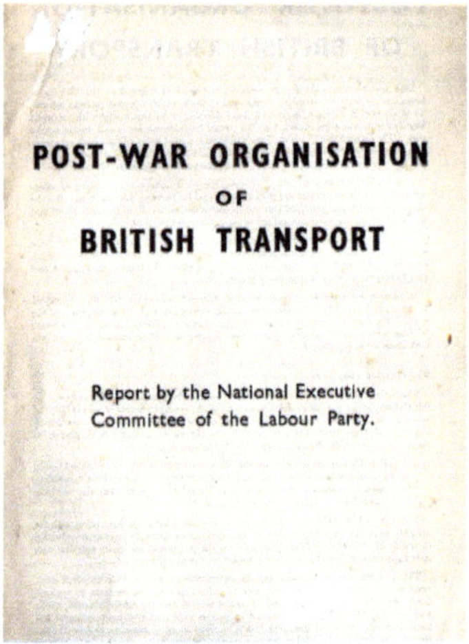
POST-WAR ORGANISATION
OF
BRITISH TRANSPORT

Report by the National Executive
Committee of the Labour Party.

Labour Party pamphlet 1943
(author collection)

It concluded that competition between road and rail was self-defeating and would never resolve itself through market forces, but that regulation would prove problematic because most road haulage was in the hands of a very large number of small businesses. The book focused on costs and profit, and ignored the social benefits of railway systems, but was accurate in its predictions of where competition would lead, as will be seen in a later chapter discussing the Beeching branch line closures of the 1960s. Walker concluded his survey and analysis with the suggestion that the coming public railway and distance road haulage monopoly should operate a system of charges based on conveyance expense rather than cargo value. It accepted that small businesses, whose scope would now be limited under the nationalisation plans, would never agree to work together with a voluntary system of charges, and that they would in effect undermine the larger monopolies with their public service obligations, through free competition.[10]

The Big Four railway companies could read the writing on the wall but, thanks to the guaranteed annual payments from Government, were reaping substantial profits across the whole system for the first time in many years, and were not about to abandon the interests of their shareholders. In 1942 they set up a commission to look at the future working of the railways, and became increasingly strident in their determination to maintain private monopolies. Said one former General Manager at a 1944 Engineering Industries Association function:

> 'Nationalisation would mean that a horde of black-coated, stripe-trousered young men, full of theory, would be playing at trains or lorries when their knowledge of transport was limited to the cost of a season ticket from their home town to Whitehall.'[11]

Between 1943 and the decision to nationalise by the post-war Government, they issued a stream of pamphlets and books that both celebrated the achievements of their staff in wartime, but also discussed the advantages of private enterprise.[12]

Conditions for railway employees in the steam age were dreadful by today's standards, and wartime pressures only added to this. This was especially the case with staff who prepared, drove and serviced locomotives. Unlike the trains of today where a single driver is backed up by maintenance staff who are not required to service motive power at the start and end of every shift, steam locomotives were highly labour intensive. The driver was at the top of a hierarchy of employees who undertook dirty, unhealthy work to prepare an engine for its daily duties and put it to bed at the end of a shift. These involved various grades of cleaner, and other staff all of whom might one day become footplate staff and even drivers. Most of these jobs involved hard physical work, and for the youngest, and those stuck at the bottom of the ladder, involved working inside locomotives scraping off the detritus of incinerated coal. William Hilton (who later became a Member of Parliament) joined the LMS in Ardrossan engine shed as a teenager in 1942. One of 300 staff, he describes the round-the-clock shifts and long hours working on (and inside) worn out locomotives hauling coal and goods, sometimes at some distance from home involving overnight stays. Here he describes the changes wrought by war:

> 'The War brought great change – almost traumatic. It provided full employment and higher earnings for most people. The railway driver no longer had a privileged position. Relatively speaking his status in society had greatly deteriorated. After 1939 this decline was also echoed in his working conditions. His fireman was no longer a subordinate; no more polished engines maintained in first class condition by the fitters. Wartime demands, and shortages of spares and labour, meant that locomotives had to be kept on the road short of positive failure.' [13]

The hierarchy Hilton describes had suited the companies as it divided employees, with drivers enjoying a superior position even though all had started as cleaners. The war brought rapid promotion alongside this reduction in status for those at the top of the locomotive end of the system, without whom nothing else functioned.

The Role of Education in the Army

After the defeat at Dunkirk the War Office became concerned about the state of morale and the situation in the army in general terms, where class, privilege and public-school education, rather than military and leadership skill, separated officers from other ranks. The feeling was that the war aims of the officer class might be different from those they commanded. Officer selection and promotion interviews were often based on questions such as 'What does your father do?' and 'What pack do you hunt with?'. From 1941, in the face of criticism, selection began to be based on psychological testing. Officers were now expected to lead by example and skill and attend to the welfare needs of ordinary soldiers.

Alongside improvements in officer selection, attention turned to army education. The Army Education Corps in 1940 did little more than provide map reading courses and had long since ceased to have the resources to undertake more general education. Now, there was a realisation amongst more enlightened military leaders that morale and education about the circumstances of the war, and hopes for the peace that would follow, were inextricably linked, and that this needed attention. This, it was hoped, would begin to address plummeting morale as the war continued to go badly for the British through 1941. This was a people's war and the army needed the same force of ideological motivation found within Cromwell's New Model Army during the English Civil War of the 17th century.[14]

The result of War Office deliberation was the rolling out of formal and systematic army education that would be regular and compulsory for all troops, delivered by their own officers through discussion groups in each unit. Material would be provided by the newly established *Army Bureau of Current Affairs* (ABCA), in the form of regular short pamphlets outlining wartime military issues, interspersed with more general current affairs matters. The aim was to promote open discussion around

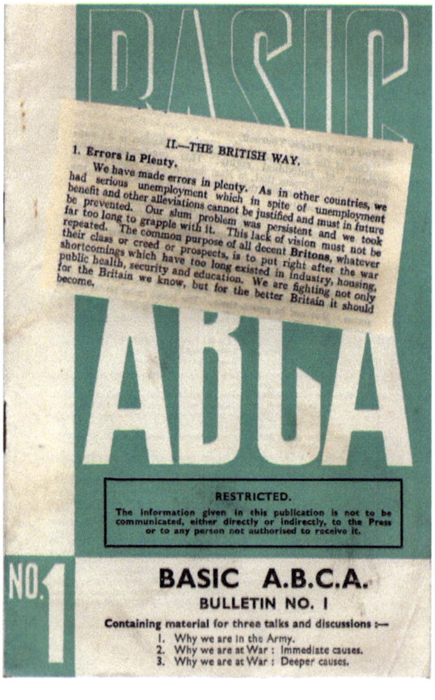

RIGHT: ABCA 1942 pamphlet No 1 and content urging interest in shaping the post-war Britain (author collection)

OPPOSITE RIGHT: Front and rear cover of ABCA pamphlet No 27 from 1942, urging interest in town planning after the war (author collection)

the whys and wherefores of warfare, and the nature of society at home and abroad. Alongside the ABCA material, sessions would also be held to discuss citizenship matters through material on *The British Way and Purpose* (BWP). A great deal of time and effort, much of it the work of teachers and educationalists recruited into the army, went into preparing the material and it was influenced, beyond doubt, by those with socialist and communist sympathies. At its height in 1945 the Army Education Corps had 2,218 officers and NCOs, many of whom were committed to the idea of a people's peace. After Russia's entry into the war on the allied side in June 1941, such left-wing influence was less contentious, although still regarded with great suspicion if not outright opposition by traditional right-wing establishment leaning officers and Conservative politicians, including Winston Churchill. This, they argued, would lead to political agitation and criticism of the leadership within the army. Major General Willans, who had devised the ABCA scheme, defended his actions by arguing that both these things were happening anyway, and his scheme would bring such discussions into the open and under control.[15]

Whilst control was exercised as the scheme got underway, some of the outcomes were not quite as expected by the army authorities – although there is evidence that they matched the mood of the times. Certainly, the opportunity for discussion which guidance suggested should be 'informal and participatory'[16],

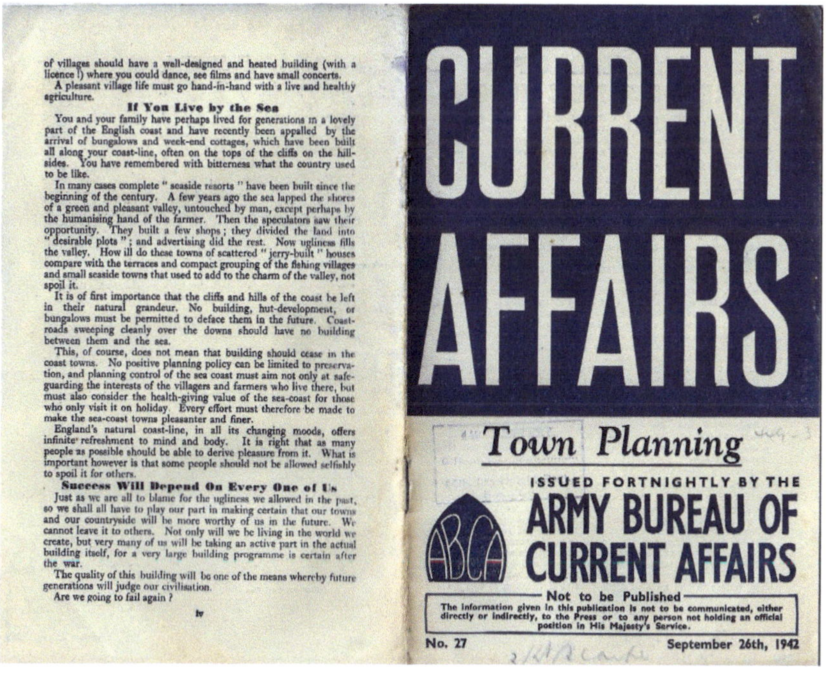
of villages should have a well-designed and heated building (with a licence !) where you could dance, see films and have small concerts.

A pleasant village life must go hand-in-hand with a live and healthy agriculture.

If You Live by the Sea

You and your family have perhaps lived for generations in a lovely part of the English coast and have recently been appalled by the arrival of bungalows and week-end cottages, which have been built all along your coast-line, often on the tops of the cliffs on the hill-sides. You have remembered with bitterness what the country used to be like.

In many cases complete " seaside resorts " have been built since the beginning of the century. A few years ago the sea lapped the shores of a green and pleasant valley, untouched by man, except perhaps by the humanising hand of the farmer. Then the speculators saw their opportunity. They built a few shops ; they divided the land into " desirable plots " ; and advertising did the rest. Now ugliness fills the valley. How ill do these towns of scattered " jerry-built " houses compare with the terraces and compact grouping of the fishing villages and small seaside towns that used to add to the charm of the valley, not spoil it.

It is of first importance that the cliffs and hills of the coast be left in their natural grandeur. No building, hut-development, or bungalows must be permitted to deface them in the future. Coast-roads sweeping cleanly over the downs should have no building between them and the sea.

This, of course, does not mean that building should cease in the coast towns. No positive planning policy can be limited to preservation, and planning control of the sea coast must aim not only at safeguarding the interests of the villagers and farmers who live there, but must also consider the health-giving value of the sea-coast for those who only visit it on holiday. Every effort must therefore be made to make the sea-coast towns pleasanter and finer.

England's natural coast-line, in all its changing moods, offers infinite refreshment to mind and body. It is right that as many people as possible should be able to derive pleasure from it. What is important however is that some people should not be allowed selfishly to spoil it for others.

Success Will Depend On Every One of Us

Just as we are all to blame for the ugliness we allowed in the past, so we shall all have to play our part in making certain that our towns and our countryside will be more worthy of us in the future. We cannot leave it to others. Not only will we be living in the world we create, but very many of us will be taking an active part in the actual building itself, for a very large building programme is certain after the war.

The quality of this building will be one of the means whereby future generations will judge our civilisation.

Are we going to fail again ?

iv

CURRENT AFFAIRS

Town Planning

ISSUED FORTNIGHTLY BY THE

ARMY BUREAU OF CURRENT AFFAIRS

ABCA

Not to be Published

The information given in this publication is not to be communicated, either directly or indirectly, to the Press or to any person not holding an official position in His Majesty's Service.

No. 27 September 26th, 1942

was seized upon by left-wing activists who were now in the army. In the North Africa/Middle East/Mediterranean theatre, where there were significant numbers of troops between 1941 and 1943, this took off with volunteer tutors from exactly that background. Lively discussion sessions ensued on what soldiers might expect after the war – these being stimulated by ABCA material when that began to arrive in 1942–3.[17] The 1942 Beveridge Report (see below) outlining proposals for National Insurance was summarised in an ABCA pamphlet, although its first edition was withdrawn due to objections by Churchill – ostensibly because it had not been agreed by parliament. The later edition included a cautionary note from the Prime Minister who, critics observed, did not want his troops fighting a war to establish socialism.[18] The report set out to tackle the issues (the Five Giants noted in the introduction to this book) that had blighted the lives of so many people prior to the war, and the idea that soldiers were fighting for victory, as quickly as possible and to change society for the better, took hold.

A total of 117 ABCA Current Affairs discussion booklets were issued between September 1941 and 1946 covering such diverse topics as the place of women in society, town planning, Scottish nationhood, health, community centres, youth clubs, schools, world hunger, art and housing. *When the Lights Go On* (July 1943), written by the Director of ABCA, was surely one of their most radical offerings. It asked discussion leaders to invite their group to lead on what sort of country might follow the war, the work ex-soldiers might do, the shape of the country they might live in (including industry and infrastructure like railways) and the voluntary effort they might contribute to make for better communities. It explicitly contrasted this with the situation prior to the war – unemployment and want. In its proposition that troops should consider their role as future citizens as a motivator to fight for victory, it pointed directly to the example of Cromwell's New Model Army.[19] The series continued after the war was over until the end of 1946, millions of service men and women still in the process of being demobilised. One of the last was titled *Everybody's Business – Local Government in England* and explained their workings and financial arrangements, as well as their significance for the organisation (at that time) of local services including transport and health. It encouraged interest and informed involvement.[20]

The BWP series focused more closely on the system of government in Britain: its detailed programme included foreign relations, empire, British history and the growth of national character, taxation, health, education, the

Christian religion, family life, trade unions, town planning and land use, and a myriad of other topics. The aim was to create an informed citizenship able to make choices for the future – and so spur enthusiastic fighting for victory. As with ABCA there was an assumption that change was required if the country was not to default to the pre-war situation. However, at least one of its subjects thought that the material offered discussion '…based on the alleged eternity and sacredness of our present social system.'[21]

Naturally enough, army education discussion using ABCA and BWP materials would have varied considerably, depending on the interest of the officers and NCOs leading them, and on the enthusiasm of the audience. However, there is evidence that some very left-leaning political discussions ensued, the most well-known being the 1943 'Cairo Parliament'. Initially backed by local Army Education staff, a 'parliament' was established in December 1943 that was open to all, attracting attendances of many hundreds of service men and women of all ranks. Debates along quasi-party lines (so as not to breach service regulations on political activity) and votes on issues such as nationalisation resulted in favourable support for socialist solutions to post-war issues. The peak of the parliament's activity, in February 1944, was a husting followed by a mock general election, in which Labour won an absolute majority from the 400 present. The next meeting, with a cabinet having been formed by 'Labour' under 'Prime Minister' Private Harry Solomons, presented a 'King's Speech' proposing policies such as nationalisation of the banks, four million new homes over ten years, increased pensions and equality of opportunity in education. It was at this point that the military leadership in the area intervened and imposed restrictions on the 'parliament's' activity that led to its demise.[22]

There were other such discussion groups elsewhere including one involving troops from the Worcester Regiment in 1944, who were fresh from the bitter battles of Kohima and Imphal in north-west India. Another debate about what to do with Germany after the war was reported to have taken place whilst troops sheltered from a mortar attack in north-west Europe after D Day.[23] As we shall see in the next chapter there was a belief widely held that ABCA and BWP discussions resulted in the growth of socialist beliefs in the services and contributed to Labour's win in 1945. However, the reality might be that the experience of total war, united purpose and wish for change when it was all over had pushed people to the left whether or not they had been given official opportunity to discuss issues collectively. [24]

Planning the Post-War Britain

As early as December 1939 a popular book on town planning suggested:

> ...there will be a great eagerness, when peace comes, to make a beginning towards a new and better way of life. There was such an idealism, though one tinged with bitterness, in 1919... The plans for the new beginnings then were all too feeble and unconsidered, the objectives too hazy and vague... we got only muddle and mess and disillusionment again. It will be infinitely tragic if that happened again. But it will happen unless we strongly determine that it shall not, unless we get the basis of our future planning policy firmly settled now, unless we have our minds crystal clear as to what we have to aim at, and how we can best achieve it. [25]

The spirit of such professional admonition was taken up by ABCA with its 1942 pamphlet *Town Planning*. This railed against ugliness and thoughtless development, and urged soldiers to become engaged in shaping the communities they would rebuild and live in after the war.[26] Planning for the peace took place in multiple settings, alongside and prompting government thinking and action. All this was spurred on by the welcome given to the Beveridge Report published in December 1942, itself the result of lobbying by the trade unions early the previous year. This addressed concerns about the inadequacy of the existing social insurance structure and laid out proposals for a new scheme that would herald a welfare state. Its publication heralded a widespread belief that the awfulness of life for many in the 1930s would not be repeated.

Glasgow represented some of the worst aspects of pre-war Britain: high density sub-standard housing, structural unemployment and a declining industrial base centred on shipbuilding. It also had a strong trade union and socialist tradition that encouraged both resistance and forward-looking political activity to build for better futures. The war brought relative prosperity – the shipyards in the region, almost silent pre-war, built almost 2,000 ships and repaired many more. In the 1922 general election Labour and Independent Labour Party candidates won 10 of the 15 parliamentary seats and this trend continued for the rest of the century. Glasgow Corporation had become Labour controlled in 1933 when a sectarian split in its main opposition saw it gain control and this stayed in place, with a short break in the late 1940s, until recent times. Although politics in theory had been set aside for the duration of the war as the main parties went into coalition, the fusion of the city's history and the resumption of hope and aspiration, alongside jobs, were bound to have some effect on post-war planning to improve life for Glasgow's citizens.

The period from the First World War until the General Strike of 1926 saw the emergence of 'Red Clydeside' in Glasgow, an era of anti-war sentiment and industrial and social unrest out of which developed an influential working-class leadership. Names that have gone down in history include Helen Crauford, Mary Barbour, John MacLean, Willie Gallagher, James Maxton, David Kirkwood and Thomas Johnston. Maxton, an Independent Labour Party MP was still active in the war years, in declining health but true to his socialist principles, as was Gallagher, by now a Communist MP in Fife. Johnston, born in 1881, had mellowed over time, having been first elected to parliament in 1922, subsequently losing his seat on several occasions, but from 1935, member for Stirling and Clackmannan. Johnston was appointed Secretary of State for Scotland by Churchill in 1940 and held that position until May 1945. Given almost dictatorial powers, he used the position to advance his own ideas about wealth redistribution, public ownership and control. In 1942 he established a Scottish Council of Industry that won so many contracts to Scotland within its first six months that government production increased by over 250 per cent.[27]

Johnston's most notable achievement was to establish hydro-electric schemes in the Highlands of Scotland in an effort to bring them into the modern age and reverse some of the decline that had taken place over the previous 150 years of clearances and emigration. The publication in 1943 of the *Report of the Committee on Hydro-Electric Development in Scotland*[28] led to a significant and unopposed nationalisation of the power supply over Scotland's Highland land mass. This not only outlined massive investment in hydro-power creation and the jobs that would come with it but took into one central public body numerous local schemes. The *North of Scotland Hydro-Electric Board* came into being in 1945, and Johnston left Government to become its first chairperson, overseeing the start of three decades worth of work to create an extensive network of dams and power stations and a distribution grid. Johnston worked hard to woo landowners (who were compensated for purchase of their land) but was clear in the report that the public interest overrode private interests, and they could not oppose his plans. These wartime proposals were exactly as Johnston and his socialist supporters intended, and their significance will not have been lost on others seeking similar solutions to social and economic problems when the war ended.

Electricity supply across the UK was itself recommended for nationalisation in the 1942 Jowitt report. This rejected the differing types of supply from multiple private companies that had accompanied the growth of the industry and its acceptance as a basic utility. As such, it eventually replaced coal as the principal energy source, with implications for the railways as we shall see in a later chapter.

One unexpected and little-known example of the push for nationalisation came from Sir Frank Whittle, the inventor of the turbojet engine, who urged Sir Stafford Cripps, Minister of Aircraft Production, to take over his invention for the state. The Government, he argued, had provided most of the investment anyway. This was not to happen, and the profits made from successful British export post-war went into the pockets of private shareholders.[29] However, change was in the wind even before the war was over, evident in the passing of the 1944 Education Act. This established free secondary level education for all, including within grammar and technical schools, and raised the school leaving age to 15. Meanwhile, work on addressing the rest of Beveridge's Five Giants was also underway with a committee to look at full employment set up in 1943, and a White Paper to create a *National Health Service* presented in 1944.

The 1942 Beveridge Report (along with a summary version) sold in its hundreds of thousands within weeks of publication; similar encouragement of detailed public discussion in other areas about the future, although very much in the air, was not present outside the restricted services circles of ABCA and BWP until the 1945 general election. One of the few MOI booklets not looking at armed forces progress, *Transport Goes to War*, published in 1942, discussed the heroic efforts of staff across inland transport sectors, but made no mention of their future.

In Glasgow, however, planning for the city was underway from as early as 1942 with the formation of the *Post-War Planning Committee* in October of that year. In conjunction with the Highways and Planning Committee, a report was commissioned in September 1943, and eventually published as the *Corporation of the City of Glasgow First Planning Report* in March 1945 (known widely as the *Bruce Report* after its principal author Robert Bruce, the city's Chief Engineer).[30] As immediate financial considerations were not required, the Bruce Report was far-reaching and ambitious, and together with the Abercrombie Report (described below), influenced city reconstruction for the next 30 years. A main and pressing concern was housing and slum clearance: thousands of homes occupied by families were one- and two-bedroomed tenement flats lacking baths and their own toilets. Both reports contained major proposals, later adopted, to build peripheral housing schemes to house people cleared from the overpopulated central districts of Glasgow. Bruce also made way for Glasgow's 1960s extensive development of high-rise flats – an opportunist response to the lack of availability of space around the city to build housing – a matter that will be returned to later in this book. The City Centre would be completely reconstructed with existing buildings, many of historical value, torn down to make way for 'a beautiful and healthy city': amongst the victims would be the City Chambers, Kelvingrove Art Gallery and Museum, Glasgow Royal Infirmary and Charles Rennie McIntosh's Glasgow College of Art.

In terms of transport, the Bruce Report placed an emphasis on road building and motorway construction linking rail, tram and bus services in modernesque hubs around the city's new arterial ring roads. Viewing the development of railway networks in the city as piecemeal and duplicative, the report suggested that the four main terminals (Central Station, Queen Street Station, St Enoch Station and Buchanan Street Station) should be demolished and two new termini built: the South Station would be south of the Clyde in the Cumberland Street area and the North terminus in the vicinity of Buchanan Street and Queen Street Stations. A new underground line would run from Pollokshields East station on the Cathcart Circle line, to the North terminus, this effectively joining the two termini. His main proposals for the suburban rail system involved replacing the tram network with an electric light rail system round the city, some of which would run along the central reservation of the new arterial road system. The report was not published with a view to immediate implementation: it was envisaged as a blueprint for the post-war future.

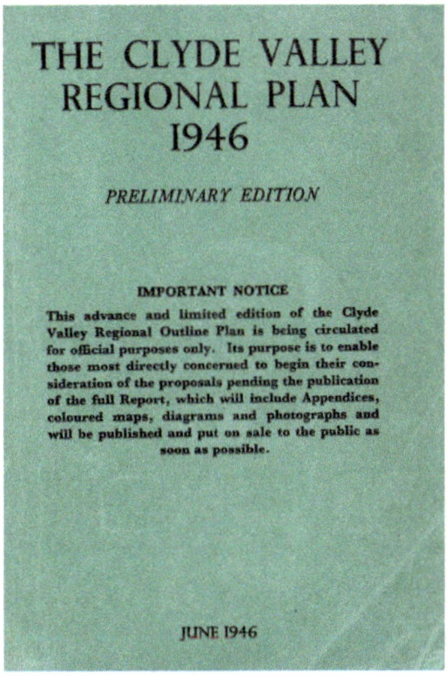
THE CLYDE VALLEY REGIONAL PLAN 1946

PRELIMINARY EDITION

IMPORTANT NOTICE

This advance and limited edition of the Clyde Valley Regional Outline Plan is being circulated for official purposes only. Its purpose is to enable those most directly concerned to begin their consideration of the proposals pending the publication of the full Report, which will include Appendices, coloured maps, diagrams and photographs and will be published and put on sale to the public as soon as possible.

JUNE 1946

Cover of the Abercrombie Report 1946 (author collection)

The Bruce Report was soon followed in 1946 by the *Clyde Valley Regional Plan*, usually known as the *Abercrombie Report*, after its main author, well-known UK-wide planning consultant Sir Patrick Abercrombie.[31] Whilst Bruce was a product of the city and its culture and worked in Glasgow, Abercrombie was appointed by the Government and an outsider – reflecting planning tensions that were to feature in the years to come.[32] The Committee under Abercrombie included representatives from Glasgow Corporation and the 13 large burghs and five county councils surrounding the city and within the Clyde basin area of the West of Scotland. Their work began in February 1944. In quite extensive detail (unlike the quite slim volume comprising the Bruce Report) its 700 pages surveyed the physical, social and economic structure of the region, before making recommendations about the future. Its scope was certainly impressive, looking at all aspects of society from housing and employment, to public utilities, population distribution leisure and recreation.

The Abercrombie Report, too, acknowledged the piecemeal development and duplication of the rail network in the region, and the poor location for historical reasons of some stations and goods yards, suggesting that major improvement was necessary. In contrast (and in effect opposition) to the Bruce Report, it suggested that Glasgow's slum clearance programme should involve dispersal of the displaced population to new towns (envisaging the development of East Kilbride, Cumbernauld, Erskine and Irvine). Whilst seeing road as the primary transport means for the new towns, rail should also be made available for rapid passenger access to Glasgow. In line with this, the report suggested electrification of lines: those out to Paisley, Greenock, Gourock and Wemyss Bay; the Cathcart Circle; Bridgeton Cross to Balloch and Helensburgh; Bridgeton Cross to Motherwell, Hamilton and Rutherglen; and Glasgow to Ayrshire (Ayr and Ardrossan) via Paisley. Noting the historical location of the four mainline termini in Glasgow, the report rejected the Bruce Report solution, and suggested that Central and St. Enoch Stations should be redeveloped to take traffic from the south and west, and north and east respectively. The report stopped short of proposals involving new lines, but advised that further work was needed to redesign the system and that the poor state of the network in the region at war's end might spur such opportunity. This did not happen for wider economic reasons, but, as will be seen in later pages, the Abercrombie Report pointed the way to the later schemes of British Railways, British Rail and Strathclyde Region.

It should be noted that both the Bruce and Abercrombie reports involved large-scale population movement, but there is no evidence in either that anyone affected was asked for their views – perhaps the central planning that directed the war effort pushed this to the back of minds, although it certainly ran counter to the extensive and wide-ranging discussions promoted by ABCA.

The previously referenced *When the Lights Go On*, issued in July 1943, suggested amongst other ideas voluntary community service:

> 'Stir up local interest in the problems of your town or district by joining or starting a Discussion Group (on ABCA lines), which would get local citizens to clear their minds on what needs doing or undoing in local affairs. It's no good grumbling at the Local Council, it's better to offer them systematic and constructive public opinions.'[33]

This was nothing other than a call for citizen participation in local politics – controversial at least and easily overlooked after the war was over. But at the time there was a mood for extensive change and active citizenship had in some places been encouraged. Whether this was realised in the post-war period is discussed in the next chapter.

Chapter 3

THE LABOUR GOVERNMENT AND NATIONALISATION

'The voice of Time cries to man, Advance! Time is for his advancement and improvement; for his greater worth, his greater happiness, his better life; his progress onward to that goal within its knowledge and its view, and set there, in the period when Time and He began. Ages of darkness, wickedness, and violence, have come and gone – millions uncountable, have suffered, lived, and died – to point the way before him. Who seeks to turn him back, or stay him on his course, arrests a mighty engine which will strike the meddler dead; and be the fiercer and the wilder, ever, for its momentary check!'[1]

Introduction

The words above are quoted from Charles Dickens' *The Chimes* and prefaced the 1945 Bruce Report that laid out post-war plans for the City of Glasgow, a work described briefly in the last chapter. The principal author, Robert Bruce, who was Glasgow Corporation's Chief Engineer, outlined a vision for the city that captured the mood of the times, so his choice of words was deliberate and profound. However, as we shall see, the report's realisation, along with that of the Abercrombie Report, proved more than either resources or favour would allow in the immediate post-war period. Changes, however, did take place that were far-reaching, but whether and to what extent they affected ordinary people and the systems, like railways, they worked on or used every day, are the subject of this chapter.

The 1945 General Election

The May 1945 victory over Nazi Germany was quickly followed by discussion within the coalition cabinet about a general election – the first in ten years as the 1940 election was postponed due to the war, with a coalition cabinet led by Winston Churchill in power from May 1940 onwards. Churchill enjoyed popular support as an inspirational war leader, but the trends described in the last chapter gave momentum to a challenge from Labour to what his supporters expected would be his reward in peacetime. Labour wanted to put off the election until later in the year, but Churchill won the argument to bring it forward to July, with a quick campaign, a normal election at home, and then a three-week wait for service votes from overseas to come in and be counted. Although many servicemen and women overseas did not vote, those cast amounted to nearly two million votes, divided between those cast by proxy and those directly from outside the country. Labour's manifesto *Let Us Face the Future* promised the implementation of the Beveridge Report, a National Health Service and control of the economy through extensive public ownership.

Churchill's response, captured in a party political broadcast on 4 June 1945, characterised Labour's socialism as heralding 'tyrannical' restraint on freedom to 'breathe freely' and 'some form of Gestapo' to control public criticism of its centralisation of power. Clement Attlee, the Labour leader, replied the following day, pointing out the benefits of long-established democratic socialism in Australia, New Zealand and the Scandinavian countries, and their freedom, and suggested that Churchill was acting as a mouthpiece for the anti-socialist newspaper magnate Lord Beaverbrook.[2]

Trafalgar Square London celebrations 8 May 1945 (*Daily Mirror*: Wikimedia Creative Commons)

A Conservative win was widely expected but the sparsely available polling showed otherwise, and when the result was announced at the end of July, Labour had won a landslide win with a parliamentary majority of 146 seats. Clearly people wanted change and, when it came to the peace, Churchill was remembered for his role in the Tory Government of the 1930s, rather than his war leadership. There was resentment too that he should be held up as winning the war when victory was the result of collective effort

Labour Party Election Poster June 1945 (© Peoples History Museum)

UK election ballot papers Hamburg Germany July 1945 (IWM Wikimedia Creative Commons)

and sacrifice. In some places the election took place amidst feverish excitement on the Labour side and huge public meetings; in others there was less enthusiasm and more scepticism about change.[3] The extent to which such ABCA and BWP discussions in the services affected Labour's win has been debated since. Churchill certainly held this opinion, as did many in his Conservative Party and those of their class in the military hierarchy. They shared a feeling that Churchill had deserved victory as a reward for his wartime leadership, and that socialists had snubbed him through their activities in army education. Subversion, it seemed, provided the explanation for such insubordination, rather than a popular will that there should be no return to the 1930s version of Britain apparently represented by his party, with their opposition to the popular measures Labour understood people craved after six years of sacrifice.[4] Labour's win heralded the most radically left-wing manifesto and most radical government ever elected in British history.

Labour's manifesto, which had sold some 500,000 copies, committed to a range of policy measures to ensure full employment, fair wages, new housing, social security and health and well-being. This would involve Government control over planning, land use and industrial growth. Public ownership of the land was invoked as a longer-term goal, but in the meantime there was an immediate promise of nationalisation across the economy including the Bank of England. As well as coal and other fuel sources, and steel and iron production, this involved all aspects of inland transport – railways, roads, canals and air transport – in order to achieve unification and co-ordination.[5] Coal and rail were by far the biggest employers involved, accounting for the vast majority of the 2 million workers whose jobs would now become part of the public sector. However, the idea of nationalisation of industries that were already under effective Government control due to wartime meant less to voters than the promise of a welfare state and a free health service.

Rail Nationalisation

Labour had committed to nationalisation of the railway system in its 1945 Manifesto, but there was debate about how this might be enacted, with pressure from the left, including within the trade unions, for worker's control. Another voice urging nationalisation for the railways, with particular application in Scotland, came from the Scottish National Party (SNP) after the general election in 1945. Although a relatively small party, the SNP had won a by-election in Motherwell in April 1945, only to lose it in the Labour landslide a few months later. The pamphlet *How Scotland Lost Her Railways* spent most of its pages berating the 1923 grouping for its loss of Scottish identity in the system and blamed the London-based shareholders (backed by the London-controlled newspapers) for putting their own class interests before those of the people of Scotland. The author, Archie Lamont, put forward evidence that the LNER and LMS companies, with trade union collusion, had moved railway wagon, carriage and locomotive production south to England, and that the private Glasgow builders had been forced to look to foreign markets. He pointed to the fact that 44 per cent of railway employees had lost their jobs in the inter-war period, whilst the figure for England was 20 per cent. His arguments were traditional nationalist ones, and he saw the railway companies as being in favour of Scotland's diminished place in the United Kingdom. He argued that this could only be avoided through Labour's plans for nationalisation – by giving Scotland full control of its railways. To do otherwise would be '...as hurtful to Scotland as the manoeuvre of capitalist solidarity which lost Scotland her railways in 1922–23'.[6] Lamont did, however, include some cogent argument about electrification in Scotland and the favour given by London-based Government to schemes in the south-east of England.

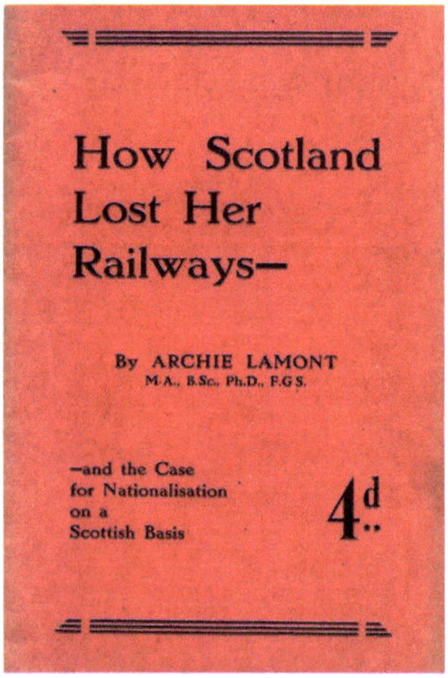

SNP pamphlet 1944 (author collection)

Despite continued opposition and lobbying by the private owners of the railway and road transport industries, the Labour Government was true to its manifesto and the Transport Act of August 1947 decreed that all internal transport systems (unless less than 25 miles in operational radius) would come into public ownership from 1 January 1948, under the overall direction of the *British Transport Commission*. The railway systems that were now to be British Railways employed around 640,000 staff, had 20,000 steam locomotives (almost one for every route mile), 56,000 coaches, over one million wagons and a variety of other moveable and fixed assets including over 7,000 horses.[7] Much of the equipment that was purchased from the Big Four companies was worn out and unfit for use and in all likelihood beyond their means to replace. In that sense nationalisation was hardly controversial: state investment was required anyway, and public ownership was already in place in the rest of Europe where systems were in an even worse state after six years of war. The stockholders did well from the compensation deal agreed and the NUR noted early on that the interest payments the former owners would receive each year (between £25 and £31 million annually) would make it difficult for them to negotiate improvements in pay and conditions for their members after 1948.[8]

Whilst the Government were now in charge, and the public nominally owned the system, there was little semblance of worker involvement in the governing of the railways. Of course, representatives of organised labour were now in government: the official NUR history comments that when their first representative, Richard Bell (then General Secretary), went into parliament as a Labour representative for the railway town of Derby in 1900, he was faced by 55 Conservative MPs who were railway directors. Now 30 railwaymen Labour MPs faced only two such directors on the opposition benches – times had changed![9] Wages were indeed held back for railway staff for many years, leading to some disillusionment, against a background of absence of any workers' control – instead complex negotiating machinery was established within which the feeling of 'us and them' persisted.

The alienation of the railway workforce is captured by the account of an idealistic young communist footplate apprentice who started work in King's Cross locomotive shed in London in 1952: he found that the older drivers were indifferent to a nationalisation that had bought them no benefit. The work was unpleasant and filthy: 'Engines are covered in grease, dirt, soot, inches thick... Prevailing picture one of utter drabness and dirt, oil,

1948 BR logo (Wikimedia Creative Commons)

grease, black soot, oily water underfoot, the air dogged with steam…' His colleagues saw no positive potential in the new arrangements and some even harked back warmly to the days of 'the company'.[10] The collective spirit that had won the 1945 election for a new future seemed far from the reality of such workplaces where work went on regardless of war and peace.

There was also an element of bureaucratic inertia with the new system: in overall charge was the British Transport Commission, with a Railway Executive, whose London offices were several miles away, responsible for operational management exercised through the six regions. There were tensions about who was in charge and setting agendas, and the result was an absence of effective planning and forward thinking by people who understood railway matters. The feeling for both staff and users was that control all seemed rather remote and out of touch with day-to-day realities.[11]

Railway passenger use also changed from that in wartime when there was regular travel across distances associated with war work and services deployment. Now the average worker travelled no more than five miles to work – double that of the 1890s, and half what it would become by the end of the century. Railways were used for commuting purposes by the middle-classes, whilst most working-class people only using them for holidays and

Coal Train Darlington May 1 1948 (© Rail-Online: Wm. Rogerson)

occasional excursions. The roads in towns and cities were the preserve of buses and bicycles; car ownership, about to boom in the 1960s, was still rare, there being two million on British roads by the end of the 1940s – middle class luxuries rather than essential necessities (by 2020 car ownership in the UK had reached 42 million).[12]

Economic Realities and Austerity

Whilst the war had brought full employment and national unity, the transition from war to peacetime production would prove problematic. Britain still had a vast empire riddled with discontented populations with a desire for freedom from colonial rule. The cheap resources that the empire had long provided had to be offset against the cost of maintaining security and government in a world where freedom from foreign exploitation and enforced rule had been amongst the values that had driven the victories against Germany and Japan. At home the call was for production of exports to bring money into the economy. Food production actually slowed down, and rationing was to continue into the 1950s.

These were years of austerity: the Government had to take out a massive loan from the USA when their wartime cheap-loan 'lend-lease' arrangement ended a few days after war's end. By this time the national debt amounted to a third of GDP (Gross Domestic Product – the value of all production and services) – debt amounting to £21 billion was only finally paid off in 2006. Millions of service people were being demobilised and had to be found work so that the situation facing many when the previous war finished in 1918 was not repeated. This was partially achieved through an exodus from the workforce of many of the women who had undertaken a variety of roles in wartime – down from 7.2 million to 5.8 million by early 1946. Housing was high on the agenda and building work absorbed many demobbed servicemen. Factories such as the Blackburn Aircraft plant in Dumbarton started producing prefabricated houses instead of warplanes. In Glasgow the situation was dire – a widely publicised report in 1946 advised that of 94,000 names on the waiting list, 40,600 were actually homeless.[13]

All of this required very careful economic planning which involved restraint on wages and restricted supply of consumer goods for many years to come, but with a pay-off in terms of a delivered national health service, a welfare state and a very low level of unemployment. Under such circumstances nationalisation might have seemed more a means to keep from collapse those services vital for rebuilding the economy, rather than opening up the opportunity for development now that income was not being used for shareholder profit. The general idea of public ownership was supported for these reasons by even some Tory politicians like Harold MacMillan. The Government went ahead with its plans and by 1950 a fifth of the economy was in public ownership. However, if the unions and their socialist supporters in the Labour Party ever thought this might involve workers' control in any of the sectors involved, or even significant involvement in management, they were to be disappointed. Whilst the Railway Executive included two leading NUR and ASLEF officials, the new boards of management included many who preferred private forms of ownership as well as a sprinkling of former generals. These bodies arguably lacked the expertise for long-term planning, a particular issue on the railways.[14]

Backwards, Forwards or Just Staying Still? – the Railways in Scotland at the Time of Nationalisation

Now burdened with fresh debt to the former shareholders, the railways were in poor shape when nationalisation came to fruition. Revenues had fallen sharply with an end to war traffic – almost halving between 1945 and 1946; coal shortages, harsh winters (especially that of 1947) and the 1948 floods also took their toll, and it took some time to sort out material and manpower shortages across industry generally as the economy reverted to peacetime production. Lack of sufficient supplies of coal, created by lack of manpower and industrial relations problems, had curtailed services in 1946 and 1947, and this had improved slightly by 1948; special trains were again put on for sporting occasions like the FA Cup Final. The annual report for the year remarked that when the new regime inspected their worn down, war damaged assets that were in arrears of maintenance and repair, and assessed what was immediately necessary to maintain services, they reached different conclusions than the private companies, and were able to make unexpected savings and improvements despite general shortages of labour and materials.[15]

The new amalgamation of the Big Four companies separated British Railways into six regions, with LNER and LMS operations in Scotland combining into the new Scottish Region. Across the country much of the equipment pre-dated the 1923 Grouping – 40 per cent of locomotives, 24 per cent of wagons and 28 per cent of coaches, and the first published plan for 1948–51 ruled out immediate replacement as impractical and unnecessary. Whilst the LNER and LMS had both planned to develop diesel-electric locomotives, and had built experimental prototypes for further evaluation, BR's chief mechanical engineer, R.A. Riddles, favoured steam locomotive development, partly because of the uncertainty surrounding oil importation at that time.

A plan to convert some steam locomotives to oil-firing was also abandoned, but these proposals had anyway made little impact in Scotland, where coal supplies were readily available from Scottish pits. Riddles, however, favoured electric power as a longer-term replacement for steam, but lacked the material means to deliver on this in the immediate future. The only mainline development in this respect in the post-war period was the Sheffield–Manchester–Wath line whose steep grades and extensive use for heavy freight had marked it out by the LNER for electrification before the war. Plans for this were now progressed and its 68 miles of route eventually opened in stages in 1954. In the meantime, under Riddles' direction, British Railways embarked almost immediately on a programme of standardised steam locomotive production to enable services to continue in the short term and immediate future.

The first passenger train on the newly electrified Manchester–Sheffield line July 1954 (© ColourRail)

In 1948 a BR committee was formed to look again at the electrification proposals made by the Weir Committee in 1931 (see Chapter 1).[16] The report appeared in 1951 and identified the 1500 Volt Direct Current (DC) overhead system (apart from the third rail system already in place in the Southern Region) as the favoured technology, and otherwise endorsed the recommendations of the earlier committee, but with no costings, timescales, or recommendations for a programme of developments.

Post-War Railway Planning in the Glasgow Area

Chapter 2 discussed wartime planning for the peace that lay in the future. The parallel and partially competing Bruce and Abercrombie reports of 1945 and 1946 respectively were intended as longer-term plans for the city, and it was their hope that they would influence, as indeed they did, further work in the future.

The first Glasgow Corporation report to follow Bruce appeared in 1948 – *A Report on the Future Development of Passenger Transport in Glasgow* whose principal author was the Corporation's Transport General Manager, E.R.L. Fitzpayne.[17] The report started life with a request from a councillor (Victor Warren) at the Transport Committee in March 1947 to look into the 'advisability' of building a linking tunnel from Buchanan St. Station to Bridge St. south of the River Clyde in the city centre, as envisaged within the Bruce proposals of a few years earlier. What became known as the *Fitzpayne Report* was the result: Councillor Warren's request was broadened in scope from simply improving the links to the east end of the city, to a city-wide scheme that included serving new housing developments. In this respect it accepted most of the Bruce proposals (including the North and South rail termini replacing the existing four) and put some flesh on them. The report sought to accommodate the rise of motor transport with the positive attributes of electrified suburban railways to provide for the future needs of the population. This extracted some suburban routes from the existing network for electrification:

Cover of the 1948 Fitzpayne Report (author collection)

- Maryhill to Queen St. Low Level via Anniesland, Partick and Charing Cross, and onwards to Bellgrove via High St. and Alexandra Park, with terminal spurs at Hyndland, Bridgeton Cross and Whiteinch;
- Central Station Low Level on a circular route through Partick, Crow Road, Kelvinside, Maryhill and Possilpark, then via Stobhill, Riddrie and Camlachie to Rutherglen and back to Central via Bridgeton Cross and Glasgow Cross; this to include the line through Botanic Gardens and Kirklee;
- The Cathcart Circle;
- The lines from Cathcart to Whitecraigs via Muirend, and Clarkston and East Kilbride from the branch at Williamwood;
- The line from Mount Florida to Central Low Level via King's Park, Carmyle, Parkhead and Bridgeton Cross.

In addition, and again as proposed by the Bruce Report, Fitzpayne proposed a new electric light-railway system fanning out from the new underground between the North and South mainline termini (which would

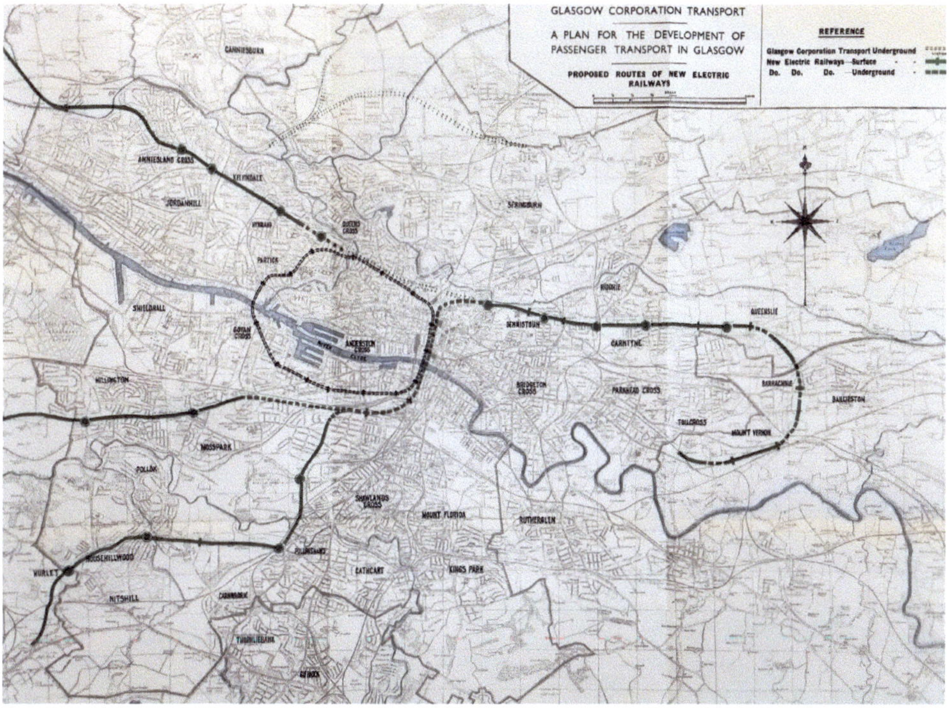

Fitzpayne Report: Plans for electrified railways in Glasgow (author collection)

accommodate the otherwise unaltered mainline routes to other parts of the country). This would follow the routes of the proposed new arterial road network and extend out to Paisley, and along Great Western Road out to the new housing development in Drumchapel, and Clydebank. It was also suggested that further development of a light railway out to Bishopbriggs from the north of Glasgow might follow. These would effectively replace trams as the 'basic means of transport within Glasgow and its environs', but a redesigned bus service would co-ordinate links with the rail-based systems to provide 'feeder', 'local', 'ring' and 'radial' services.[18] The report looked at best practice in other cities across the world, particularly at the systems in Toronto and Montreal. It made positive note of the fact that public ownership of transport could now render co-ordination between the various aspects of the network easier. It also went into some technical detail (absent in the wartime reports), including proposals for light-railway car design, and mentioned the possibility of the then new hydro-electric development as a power source. The report lacked detailed costings, but Fitzpayne countered potential costs-based criticism of his proposals by showing receipts and expenses for the existing Subway: these demonstrated that after 1942 the

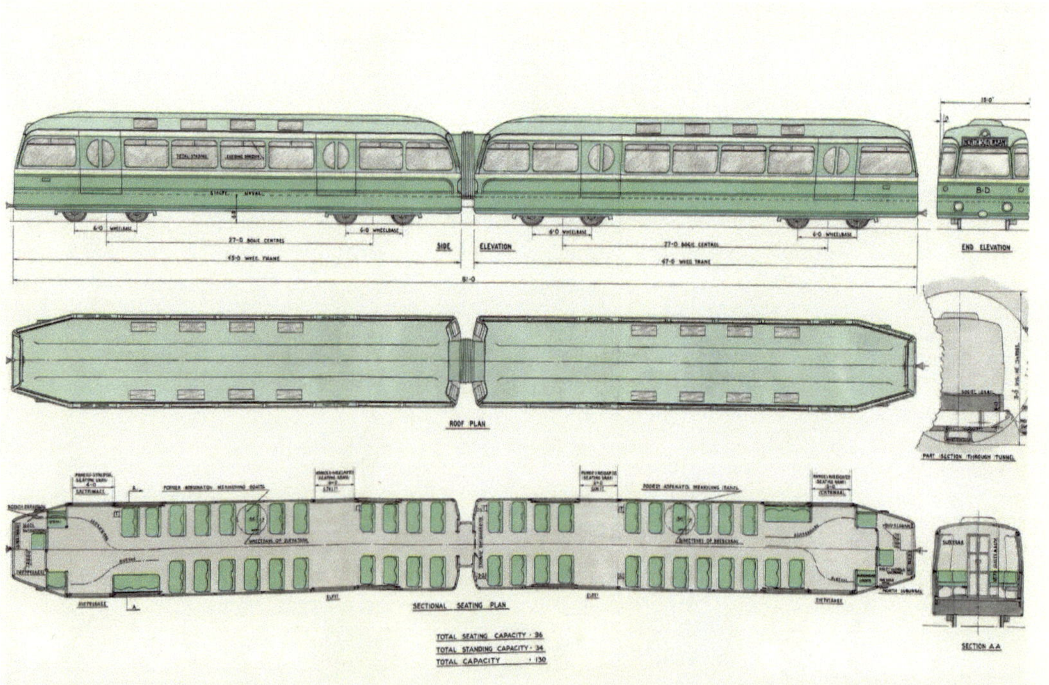

Fitzpayne Report proposed light railway car (author collection)

Fitzpayne Report proposal for the interlink at Kelvinbridge (author collection)

system had begun making a surplus that was continuing. He also suggested that removing the trams and replacing them with a light-railway system that did not share rapidly congesting roads would save lives through reducing accidents. As we shall see later in this book, his successors over the next 20 years built some new roads and motorways, and removed the trams, but unfortunately simply replaced them with buses.

Glasgow and Urban Planning

As mentioned earlier in this chapter, Glasgow's housing situation required urgent attention in the post-war period, and transport plans had to fit with the implementation of plans to clear slums and rehouse hundreds of thousands of people, rather than the city as it then was.

Whilst Glasgow officials favoured solutions that kept their people within the city boundaries, the Scottish Office of the new Labour government saw the development of new towns as being the way forward for Britain's overcrowded industrial cities with their slums and overcrowding – consolidated by the New Towns Act 1946. This debate was partially reconciled by the limited availability of building land within the city's green belt – Glasgow being a conurbation built between hills in the valley of the River Clyde. Abercrombie had proposed the relocation of over a quarter of the population (up to 300,000 people) to new towns that had yet to be built, and about the same number to peripheral housing schemes within the city. In the end this is largely what happened. Such processes led to the rise of the new towns of East Kilbride, Cumbernauld, Livingstone and Glenrothes (the last two being far from Glasgow) and the extension of existing communities in Erskine and Irvine. The peripheral housing schemes of Drumchapel, Castlemilk and Easterhouse, mainly involving four-storey tenement flats, developed in the 1950s, were eventually criticised for their lack of amenities (a deliberate policy when it came to the availability of pubs and betting shops) and isolation. So, in the 1960s Glasgow Corporation turned increasingly to high-rise flats which could house large numbers in a smaller space wherever there was land available across the city's communities. By the 1970s Glasgow had over two hundred high-rise buildings, some over twenty storeys high, more than any other comparable city in Europe. By then urban planners began to realise these were not good environments for vulnerable adults or families with children. Some were poorly designed and not built with materials for Glasgow's climate, and they in turn were condemned and replaced in the early years of the present century.[19] These tensions were to dog transport planning in the years to come and explain the absence of the linear evolvement of railway systems in the Clyde area.

There were other factors that came increasingly into play: the main one has origins in the Bruce and Abercrombie reports – namely the prime place being given to motor vehicles and the consequent development of motorways in the area. Glasgow never developed a ring-road system but a road tunnel that gave road transport access under the river in the west of the city at a time when there was still considerable shipping movement, and the M8 motorway that sliced the city in half through the centre, despite Glasgow having a lower car ownership than other British cities where there were similar developments.

The other big factor, unforeseen in the 1940s, was Glasgow's rapid de-industrialisation through the last half of the 20th century. Between 1951 and 1991, census figures show that the city's industrial workforce fell from half of all employees to 19 per cent. All the factors that had given it a prime place in the industrial revolution were in decline – its significance as a trading port, its proximity to the resources that serviced its heavy industries, and

Post-War housing Easterhouse, Glasgow (© Alamy: Tom Kidd)

these key industries themselves. Initially, lost jobs were compensated for by new opportunities such as the Linwood car factory just outside Paisley. However, by the 1980s there was nowhere for displaced workers to go: the city now gained a name for deprivation characterised by poor health (with addiction issues to the fore), high mortality, unemployment and inequality. Such social problems drew heavily on the public purse and inevitably diverted funds from infrastructure projects including those associated with railway development. Glasgow's population now fell from over one million in 1951 to less than 600,000 at the end of the century, reducing the city's income and changing, but not necessarily reducing, transport needs.

This is jumping ahead in our story, the focus at this stage being on the 1940s post-war period, when there was almost full employment and a general air of optimism that life would get better for everyone. That it didn't quite turn out like this for the rail system will be examined later.

Chapter 4

1950s TRANSPORT MODERNISATION

> 'Railways, roadways and waterways are all now essential parts of a national system of highways in the broadest sense, and in the light of the conditions obtaining and which are progressively deteriorating, it is more than time that drastic measures were taken to bring some order out of the prevailing chaos. Fair and equitable conditions must be provided as far as practicable under which the various forms of transport can function efficiently within their respective spheres and as part of a comprehensive whole. Competition in the past between undertakings operating on rail, road and water has, without doubt, fostered reduction in charges and at the same time encouraged new and improved services, but uncontrolled competition has led to much wasteful duplication of services and to a lack of co-ordination which is undesirable from a national viewpoint.'
>
> INGLIS REPORT (1951)[1]

Introduction

The early 1950s saw a change in government from Labour to Conservative but a continuation of what became known as the Post-War Consensus – a period that lasted from 1945 until the late 1970s' advent of neoliberalism. In simple terms the Tory governments of that era were committed to continued management of large sections of the public sector within a mixed economy, alongside expansion of the welfare state and NHS; this was eventually replaced by a rightwards turn to the market as regulator and arbiter of services and industry. However, one of their earliest decisions was to reverse public ownership of the potentially most profitable aspect of the 1948 nationalisations through the Transport Bill of 1953. This returned road transport to the private sector, ending co-ordination and reintroducing competition between road and rail, accelerating the decline of railway traffic so that line closures became politically inevitable and realised through the Beeching Plan of the early 1960s. The British Transport Commission now took on direct responsibility for the railways and canals, scrapping the Railway Executive. This chapter describes Britain's faltering steps towards creating a modern railway in the 1950s and how this merged with the Glasgow planners whose post-war visions were discussed in previous chapters.

The Festival of Britain 1951

At the height of the Victorian age, Britain hosted a huge exhibition in London, to showcase to the world, and the British people, the extent and magnificence of its empire and the industrial development based on its proceeds. One hundred years later the Labour government decided to hold a Festival of Britain in 1951. Whilst the centrepiece was on the Southbank in London, other exhibitions and events took place throughout Britain, '... so that this country and the world could pause to review British contributions to world civilisation in the arts of peace.'[2] The Festival was a celebration of arts and culture as well as exploration, innovation and manufacturing. The London Festival venue was surrounded by bomb sites from the war and some wondered why such an expense had been contemplated when there was so much reconstruction still to be done. However, it was popular and visited by hundreds of thousands, and regarded as a success.

The London Southbank exhibition included a transport section with a railway feature consisting of rolling stock and examples of infrastructure. This showcased steam locomotives built by Glasgow manufacturers for export. The exhibition also included a new electric locomotive for the Manchester–Sheffield–Wath line that was in the process of construction – a loco design that was by then ten years old, and both austere and functional in its lines.

British Railways (BR) produced a special large-format 50-page magazine for the festival, most of which was devoted to steam age technology and images of the railways that would have been recognisable many years previously. In their introduction BR justified not setting out to '... display

RIGHT: Royal association sells cars! Advert from the 1951 Festival of Britain official programme (author collection)

OPPOSITE LEFT: Showcasing BR: cover of *Your British Railways* 1951 (BR: author collection)

OPPOSITE RIGHT: Ambiguous back cover of *Your British Railways* 1951 (BR: author collection)

only the new and spectacular'. Its stated object, they explained '... is rather to reveal something of the complex romance of organised human effort behind those familiar things about the railways which are accepted as commonplace and taken for granted.'[3] That might have pleased some but was hardly likely to help overcome the image of a tired system – the experience of their 982 million passengers annually. In fact, just two pages were devoted to innovations in locomotive power, and these gave a brief overview of the new standard steam locomotives being built in numbers, and the very few experimental diesel and electric ones inherited from the private companies in 1948. The only new British Railways non-steam mainline loco mentioned was a gas-turbine powered experimental design – which was eventually discarded.

The focus of the Glasgow exhibition in the Kelvin Hall was on 'industrial power' and as such contained a large central section devoted to shipbuilding and railways. The latter had examples of railway engineering, models of locomotives, and two actual examples: a large 4-6-4 North British built steam engine for export to Australia, and a small industrial 0-4-0 diesel shunting engine.

Meanwhile other parts of the exhibitions (London and Glasgow) went to great lengths to show off technological innovation and the latest in design.

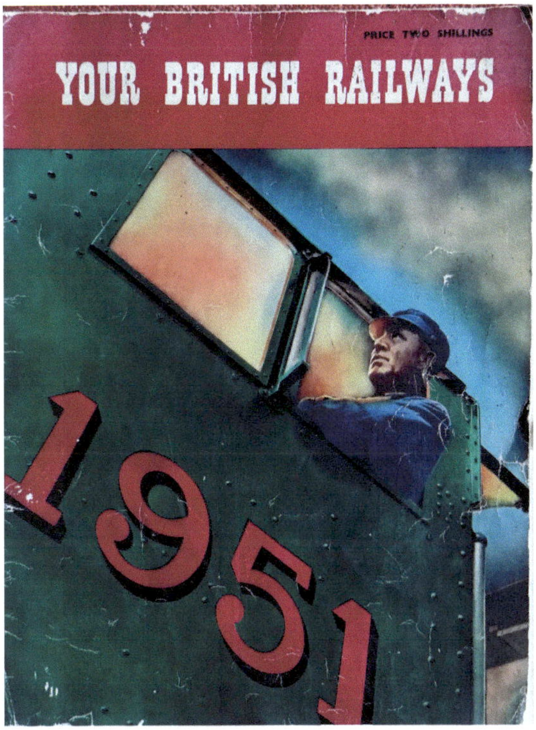

The Glasgow one featured very imaginative coal, hydro-electricity and forward-looking displays that included the possibility of nuclear power for peaceful energy production. It should be remembered that the railway magazines of 1951 were reporting from beyond our shores on a newly electrified 200-mile main line between Paris and Dijon in France, where high-speed travel was now a reality. French railways at the end of the recent war were in a far worse state than Britain's, having suffered relentless bombing by the allies when under German occupation. They had made a start with electrifying main lines before the war, as had other countries in Europe; it was only in Britain and Eastern Europe that there was still volume production of steam locomotives by the early 1950s.

The exhibition in London contained a display relating to road transport in keeping with the nationalisation of bus and road transport in 1948. This emphasised integration with rail rather than, as was about to explode over the next 20 years, a universal message extolling the freedom and necessity of car ownership and the flexibility of road haulage of goods. The car manufacturers displayed their wares, but with an eye to export rather than an expanding home market. The 1951 Festival of Britain was a celebration

TOP: The modern locomotive – from *Your British Railways* 1951 (BR: author collection)
BOTTOM: Glasgow-built loco for India displayed outside the main Festival of Britain site in London 1951 (Wikimedia Creative Commons)

of wartime survival and hope for the future. Had it taken place a few years later it might have had a very different and more competitive outlook aimed at home markets as the new decade gradually saw a raising of wages, standards of living, disposable income and expectations.

Smog and Electricity

The other factor driving progress in the early 1950s was, as mentioned above, that of air quality in the cities, especially London. Domestic and industrial coal use (and, by then, increasing vehicular traffic) had been the cause of ongoing smog (smoke and fog) since the previous century, but were particularly responsible for a crisis in the capital in December 1952. For four days, until the weather changed, London was beset by a thick fog that reduced visibility to a few yards, paralysing public transport (apart from the underground trains) and causing breathing difficulties for those venturing out. Estimates vary, but 10,000–12,000 deaths may have resulted from this episode.[4] One of the contributory factors was the recent abandonment of electric trams in favour of diesel buses. The impact of smog resulted in legislation governing domestic coal use in urban areas, and also spotlighted the necessity of clean air and the measures that might be taken to improve air quality. This included extending the National Grid so that electric power was more widely available, leading to the concept of central heating in homes, initially by electricity and

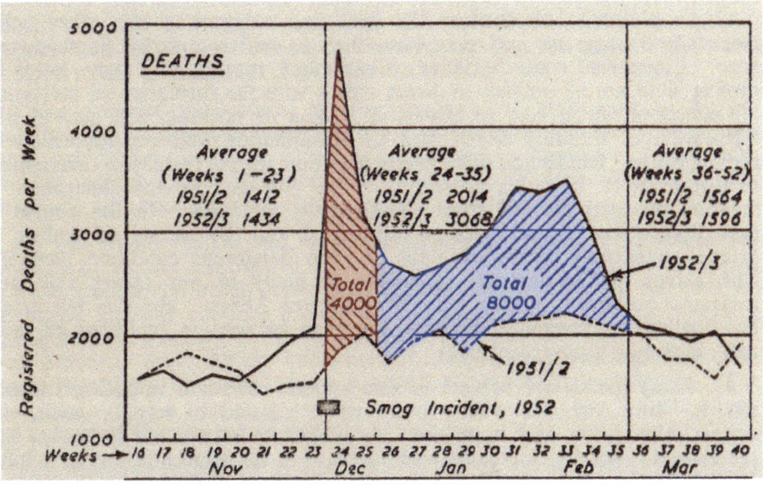

Graphic description of deaths in London from smog in 1952 (HMSO: *Committee on Air Pollution – Interim Report 1953* Wikimedia Creative Commons)

more recently by gas, but effectively ending coal as the main source of home heating. This gradual move changed the shape of the railways: it merged with plans to replace steam motive power and helped along the process of reducing the rail network, much of which was used to transport coal from the pits where it was mined, to domestic and industrial users.[5]

Inglis and Planning a Rail System for Glasgow and the Surrounding Region

Whilst the outlook portrayed by British Railways for the Festival of Britain might have been lacklustre, in Glasgow a Regional Committee chaired by Sir Robert Inglis had been organised in 1949 by the British Transport Commission (BTC) and the Secretary of State for Scotland, to look at services in the district.[6] Committee members were drawn from railway and bus managements within the jurisdiction of the BTC but, at various stages, meetings took place with local authorities and others involved in physical planning processes in the region – reaching out as far as Ayr, Helensburgh, Dunbarton and the new town of East Kilbride. Eric Fitzpayne from Glasgow Corporation was appointed to liaise with the City's own Transport Committee, but did not sit as a member. Starting from the work already undertaken within the Bruce, Abercrombie and Fitzpayne reports, Inglis and his colleagues looked widely at all forms of transport in the area alongside the plans to improve housing and develop new towns. The report's findings were better researched and explained than their predecessors and, considering the times, ambitious and forward-looking. Although it did not consult with the public, it did analyse information obtained about transport use in the Glasgow area, and used these to speculate about the outcome of improvement to systems. Its introduction, however, noted caution and the need for continued consultation with others, especially Glasgow Corporation who ran the tram and city bus systems. It was also noted that the capital investment needed to meet its vision would be affected by the '...overriding demands of the national re-armament programme'[7] – a reflection of the Korean War

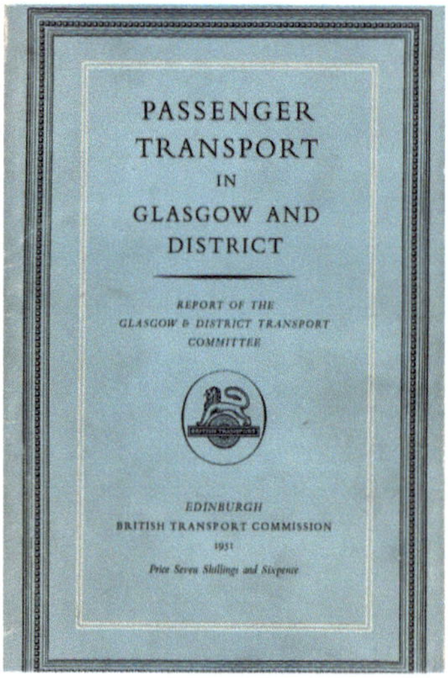
PASSENGER TRANSPORT IN GLASGOW AND DISTRICT

REPORT OF THE GLASGOW & DISTRICT TRANSPORT COMMITTEE

EDINBURGH
BRITISH TRANSPORT COMMISSION
1951
Price Seven Shillings and Sixpence

Cover of the Inglis Report 1951 (author collection)

and, behind it, ongoing Cold War and post-war decisions that Britain should be a nuclear power. However, the summary stressed that proposals were primarily based on making best use of infrastructure already in the public sector rather than outlining new systems that would require burdensome heavy capital expenditure (as arguably proposed in the Bruce and associated Fitzpayne reports). The overall mission was to reduce increasing congestion on the roads by co-ordinated and integrated services and infrastructure – all under an assumed and preferred model of public control through nationalisation.

Taking account of the recommendations of the previous reports and their feasibility or otherwise, Inglis went into detail based around reality as well as vision. The Bruce and Fitzpayne reports' premise of extensive Glasgow City Centre reconstruction, and two new railway termini, was discarded as were associated infrastructure construction such as the new underground railway connecting them, and main lines east, south and west. A more modest proposal was made that Queen Street and Buchanan Street Stations should be combined and the latter's extensive and busy goods yard be relocated – all as part of city centre redesign and street realignment. This aspect would have involved the effective closure of Queen Street Station High Level and the diversion of its traffic into an enlarged Buchanan Street terminus. Following Bruce and Fitzpayne, it would phase out trams and instead provide extensive electrification of suburban railway services, with feeder buses, and new stations serving the newly built and developing towns and communities. The report's argument for this was based on an analysis of the historical development of railways and trams in the Glasgow area and how competition had resulted in the popularisation of trams, buses and road transport, and the consequent reduction in use of railways, leading to lack of investment and decay. Electrification would reverse this trend, and mention was made of the long-closed station at Botanic Gardens in the West End of the city as an example: an architecturally attractive station building negated by a densely smoky atmosphere when trains passed, resulting in early unpopularity and eventual disuse.

Inglis also looked at population changes, both as those then being experienced, and those projected through the building of new towns and peripheral housing schemes as proposed in the long-term Abercrombie Report, and others related to housing and community that had appeared since 1946. This took account of both population dispersal out of the over-populated centre of Glasgow, and population growth generally. It considered that future transport needs could be met through better use of existing routes, some new connections and new, enlarged and re-sited stations. The report also looked at fare structures and comparisons between the different operators, demonstrating that the gap between the more expensive rail fares and the buses and trams had expanded between 1939 and 1950. Where possible (i.e. within services run by the public sector), ticket sales should be restructured, with availability of through tickets across differing modes of transport. The report envisaged more regular services through more efficient use of train sets that could be run in either direction – steam-drawn suburban services lost considerable time in turnaround and servicing.

Quoting from other reports (for example the Weir Report of 1931 mentioned in Chapter 1) and experience in Britain and abroad, Inglis calculated that immense savings could be made on the reduction in coal consumption that electrification would bring – 70,000 tons a year in the Clyde Valley area – arguing that this surplus could be exported. In the 21st century we would cite other urgent reasons for such savings. Energy requirements could partly be met through existing power resources by the gradual elimination of electric trams.

In hindsight, living in an age where there is regret about the closure of tram systems and concern at the great public expense to build new ones (such as in Edinburgh) it is easy to be critical of the recommendation to gradually replace trams with buses and trolley buses. Inglis argued that the rail electrification plan would reduce tram use, as people now living over a wider area in Glasgow and its surrounds would value rapid electric rail transport into the city centre for work or onward journeys. Trams in 1951, the report acknowledged, were at the limit of their capacity, overloaded at peak times, and contributing to road congestion in the city centre with its narrow streets (Glasgow at that time had 22 per cent of all the tramcars in Britain). In reaching this conclusion, Inglis was following others, including a 1930 Royal Commission on Transport which examined the pros and cons of trams and concluded that they were 'an obsolete form of transport...(and) cause much unnecessary congestion and considerable unnecessary danger to the public'.[8] In 1951 the Inglis recommendation did not go down well with pro-tram lobbyists: the Light Rail Transport League proposed that the tram system should be developed into a rapid transit rail network over a 20- to 25-year period along the lines suggested in the Bruce Report. This would have removed trams from roadways and avoided replacing them with more expensive buses and trolley buses. It could also have reduced future investment in a railway

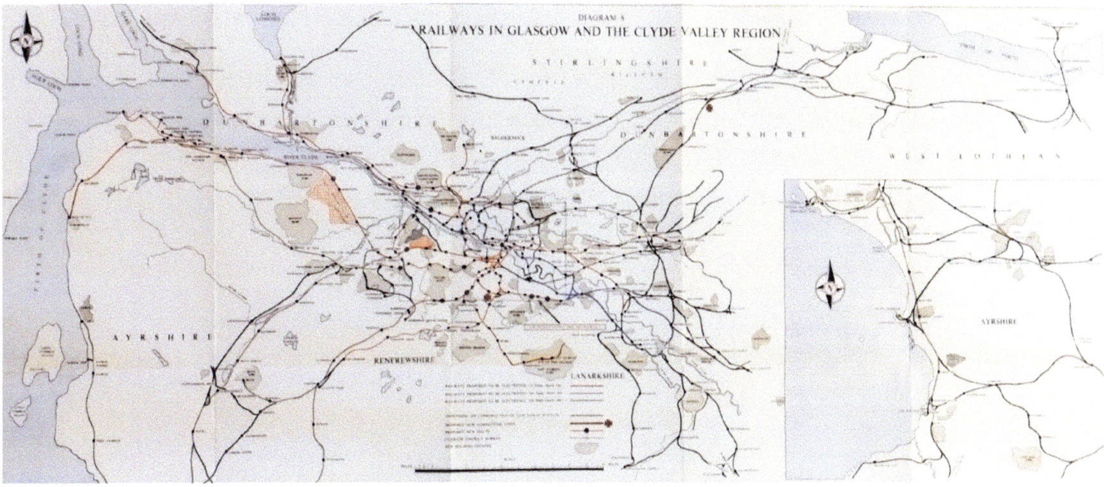

Inglis proposals for Clyde Valley railway development (author collection)

system that could not geographically meet the region's changing needs.[9] Their slim 16-page report, unlike Inglis's 81 pages, was not costed and fell by the wayside only to find resurrection in proposals put forward regularly in later years.

Inglis proposed a two-stage development. The first would involve electrification (using the then favoured 1500V DC system) on the following 153 miles of railway:

(a) Lines centred on Queen Street Low Level: Helensburgh and Balloch (via Singer and Clydebank) and Milngavie to Bridgeton Cross, Bothwell and Airdrie. Knightswood North and South Junctions to Bellgrove via Possil.

(b) Lines centred on Glasgow Central High Level: Central to· Gourock, including four tracks between Shields Junction and Paisley (Gilmour Street). Central to Cathcart Circle, Kirkhill and Uplawmoor. Wemyss Bay Junction to Wemyss Bay.

(c) Lines centred on St. Enoch: St. Enoch to Paisley West (including Union Line at Shields Road). St. Enoch to Barrhead Central. Busby Junction to East Kilbride.

(d) Extensions of the Central High Level System: Arkleston Junction to Renfrew.

This would be followed by a second stage involving 37 miles of railway:

> Lines centred on Central Low Level: Dalmuir to Motherwell via Rutherglen. Stobcross via Kirklee and Kelvinside to Partick West. Bridgeton Cross to Newton via Carmyle. Newton to Motherwell via Hamilton.

A third stage involved 118 route miles:

> (a) Rutherglen Junction to Buchanan Street, via Coatbridge and Gartcosh
>
> (b) Greenock (Princes Pier) to Paisley West
>
> (c) Ayr to Paisley (Gilmour Street) via Lochside and Milliken Park or the alternative routes between Dairy and Johnstone and Elderslie, the second one to be considered if and when traffic levels warranted
>
> (d) Largs to Kilwinning via Ardrossan
>
> (e) Kilmarnock to Barrhead, via Lugton
>
> (f) Buchanan Street to Lenzie, Kirkintilloch and Lennoxtown (conditional on Queen Street Station being combined with Buchanan Street).

These proposals were based on analysis of where pressure was greatest at that time, with the second stage being seen as a continuation of the first. The third stage would presumably follow at a future time when resources were available and proof of need demonstrated. Inglis estimated that for stages one and two, additional annual revenue from increased passenger use would amount to £200,000 (£5.5 million today) after additional expenses and interest on money borrowed for the construction cost of £14 million (£386 million today) were taken into account.[10]

One factor that might have been considered by Inglis was the ready availability of renewable electricity in Scotland by the early 1950s through the efforts of the North of Scotland Hydro-Electric Board (NSHEB) – referred to in Chapter 2. Although electrification in other mountainous regions in Scandinavia and Switzerland was proceeding apace, there was no such discussions in Scotland apart from one experiment with battery power. This involved a two-car Battery Multiple Unit on the Aberdeen to Ballater branch

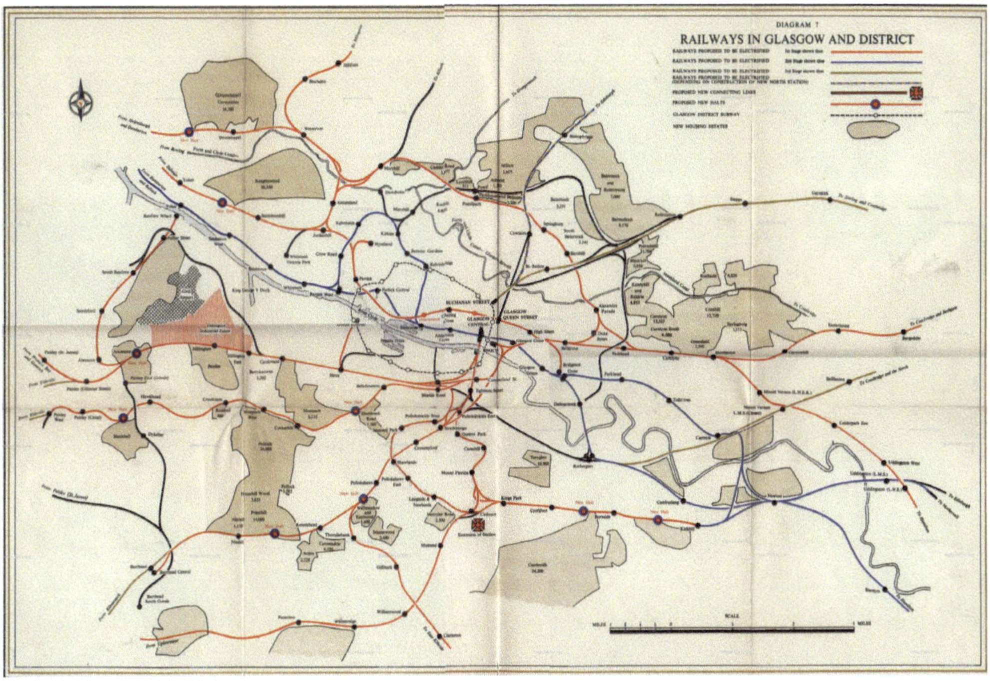

Inglis proposals for the immediate Glasgow area (author collection)

(known as the Deeside) line – a collaboration between BR and the NSHEB with services beginning in April 1958. Charging stations were constructed in Aberdeen Station and 43 miles away in Ballater as both were needed. The single BEMU, a converted standard DMU, operated alongside steam (initially) and then DMUs for several years and was considered effective; passengers enjoyed the silent and smooth operation, and drivers preferred its performance to the diesels. However, by 1962, faults had developed, and it was withdrawn from regular service, appearing elsewhere on the network in England in subsequent years. The Deeside line closed in 1966 and the drive unit eventually ended up in preservation, back on a reopened section of its original route – the Royal Deeside Railway. The battery technology was in its infancy and was not taken up elsewhere by BR, but the idea has returned in our own times with improvements in weight and performance (described in Chapter 8). In an age where most services were steam powered there are no contemporary reports of the absence of pollutive fumes from the BEMU compared to its diesel counterparts.[11]

There was hardly time to look at implementation of the Inglis Report when it was effectively torpedoed by the change of government and 1953

Battery Electric train on the Ballater Branch late 1950s (© Rail-Online)

Transport Act that reintroduced competition between road and rail, and made the co-ordination required less achievable. Not that this was crucial to undermining the report, as other forces were at work too. Although Glasgow Corporation had been involved in the Inglis Committee through Eric Fitzpayne, its 1951 City of Glasgow Development Plan ignored it, and the planners involved in the new towns of Cumbernauld and East Kilbride did so too. Fitzpayne in fact attacked Inglis in a report to the January 1952 Glasgow Corporation Transport Committee, criticising the rail biased membership of the committee, the integration method proposed (he preferred an area committee along the lines of the Clyde Navigation Trust) and the elimination of trams.[12]

Within the BTC itself, there was an absence of support: 17 separate closures of long run-down rural branch lines in Scotland took place in 1951, and with passenger use on some Glasgow suburban lines running as low as 3 per cent of capacity (the very issue Inglis set out to remedy), the Cathcart Circle and Kirkhill lines as well as those to Uplawmoor and East Kilbride were suggested for closure in a 1953 report.[13] Figures for the early years of nationalisation show that profits (or lack of them) for each of the regions was roughly commensurate with investment in modernisation. The Southern Region, with its extensive network of electrified lines, was generally profitable, whilst the Scottish Region during those years, with no electrification, made a slight profit in 1948 and 1951 only – a result of the lack of comparative investment by the LNER, LMS and GWR companies prior to nationalisation compared with the Southern Railway.

In 1953 Inglis's work was revived when the Secretary of the British Transport Commission asked the Railway Executive to look at implementation of the report. Apparently, interest was being taken by the Minister of Transport and the Secretary of State for Scotland. This resulted in a follow-up, commissioned by the Scottish Regional Board and appeared as a confidential, typed, internally circulated report in 1954.[14] The committee, under the chairmanship of BR official A. Stewart, had been asked to examine the costs and engineering requirements for the electrification of the North Clyde lines between Airdrie and Helensburgh, including the branches to Milngavie and Balloch; in the South Clyde, similar was asked in relation to the lines from Glasgow Central around the Cathcart Circle, from Cathcart East Junction to Kirkhill, and from Cathcart East Junction to Neilston High. Gone were all the other proposals in the Inglis Report, quietly cast aside by the BTC and Scottish Regional Board. However, an appendix to the report went beyond its initial remit and as a proposed Phase 2 examined the feasibility of electrification of the line from St Enoch (rather than Central) to East Kilbride, and from Central Station out to Gourock and Wemyss Bay via the line through Paisley Canal and Kilmacolm. It suggested that the former was justified because of the development of East Kilbride new town and that service improvement was now requested by the new town authority. As regards the branches to Gourock and Wemyss Bay, the report suggested that traffic figures demonstrated that these had more potential than the lines proposed in Stage 1 because of their links to ferries serving the Clyde coast.

The 30-page report summarised basic information relating to costs, engineering requirements and train set details for the Stage 1 North and South Clyde proposals, advising that this could be completed within five years. Track mileage involved was 51 for the North and 70 for the South. Ninety-one new three-car Electric Multiple Units would be required – 58 for the North and 33 for the South. Increases in comfort would obviate the need for First Class carriages, introducing an egalitarian element not found on the Southern Region suburban services. The new train sets would enable increased frequency of service due to better acceleration and reduced turnaround time. Footplate and engine cleaning staff would reduce from 150 to 60 'motormen', whilst guards and porter-guards would reduce from 51 to 48.

The report did make some brief but interesting comment on the demographics involved in both proposals – noting that the South lines would attract 'business travellers' whilst the North ones would cater predominantly for 'working people'. The report noted the development of the Castlemilk housing scheme and that this could be served by South Clyde services to Croftfoot and Burnside. Power supply would, as with Inglis, come through the 1500 V DC system and the report outlined where electricity substations would be required, and stated that engineering work to bridges and tunnels would be needed for the installation of overhead lines, but that this would not present problems. The report, like its predecessors, was clear that electrical coloured light signalling would be required throughout the system, a matter not attended to previously in the Glasgow area despite improvements elsewhere in Britain before and after nationalisation. For construction it advised that British Railways' own Signal and Telegraph staff (S&T) could undertake the signal work, its own Civil Engineering staff the engineering work, and that contractors should install the overhead line and electrical equipment.

The total cost was estimated at almost £11 million (£264 million today). Stewart concluded pessimistically that whilst revenue would increase annually by over 50 per cent at £140,000 (£3.34 million today), the overall net position of BR would be worsened by £68,000 annually because of the initial costs, increasing the financial loss these services were already encountering. In this respect he remarked that competition with road and trams would continue to reduce revenue unless a controlled and co-ordinated approach, as outlined by Inglis, was taken to reduce chaotic roads congestion into the city centre. He also concluded, as mentioned above, that more financially viable Stage 2 extensions out to the Clyde coast could offset losses incurred by the Stage 1 services.

The *Railway Magazine* reported in November 1954 that the BTC's 'Glasgow Passenger Transport Committee', presumably prompted by the internal Stewart Report, proposed a joint committee with Glasgow Corporation to discuss implementation of the Inglis proposals. Any findings of such a committee would not be binding. An area body was also sought by Fitzpayne, the city's Transport General Manager, although with more ongoing planning powers, something that was not to happen until the 1970s as we shall see in a later chapter. In the meantime, Fitzpayne's 1955–6 Annual Report to the Corporation made no mention of the threat to the city's tram services that Inglis proposed – or indeed any reference to railway electrification plans, simply listing the curtailed tram services.

British Railways Decides to 'Modernise'

In somewhat scaled down form compared to Inglis, lacking the artistry and imagination of the Bruce and Fitzpayne Reports, the often-overlooked internal Stewart Report at least put Glasgow suburban electrification back on the map, and pointed to future development in a second stage.

Meanwhile wheels were turning in the headquarters of the BTC in London. Critics of the 1948 railway nationalisation complained that assets available for modernisation within a nationalised industry had to compete with other public sector budgets – education, health etc. whilst the private companies could allocate whatever they could afford to modernisation. Proof is cited by way of the fact that the modernisations that did occur in the immediate post-nationalisation years were all as planned by the private Big Four companies e.g. the Manchester–Sheffield electrification.[15] This is not the place for this debate, but suffice to say here that previous chapters have already cited examples of the chaotic development of the railway system within a framework of competition, and how this had played out by the time the system was at last nationalised under quite unfavourable circumstances after the Second World War. These circumstances might explain the muted celebration of Britain's railway past and future at the Festival of Britain in 1951, and the modest approach taken to modernisation in these early years.

By 1955, the situation looked slightly more favourable and BR published its *Modernisation Plan* based around investment of some £1,200 million (£27.5 billion today).[16] This seems an enormous sum, but as the report explained, half of this would have been required anyway to maintain existing services. The report started out by observing:

'Railways today are not working at full efficiency, mainly owing to their past inability to attract enough capital investment to keep their physical equipment fully up to date.

This plan aims to produce a thoroughly modern system able to meet both current traffic requirements and those of the foreseeable future. It is based on the premise that its main components shall be capable of being started within five years, and completed within fifteen years.'[17]

The report focused on five basic requirements:

1. Modernisation of track, signalling and telecommunications to enable higher speeds.

2. The replacement of steam traction by several thousand diesel and electric locomotives.

3. The modernisation of passenger carriage stock involving replacement by new carriages, and electric and diesel multiple units. The associated modernisation of numerous stations and parcels depots.

4. The remodelling of freight services, involving wagon replacement by continuously braked and in some cases larger vehicles, a reduction of the number of marshalling yards and their re-siting and modernisation.
5. Some funding to be given to modernisation of sundry items such as office equipment, staff welfare, ports, and research and development.

The report considered that the benefits from this expenditure would improve public confidence and use of the system, reinvigorate staff morale and reduce the problem of road congestion. It would increase revenue by £85 million a year (nearly £2 billion today) within a context of an overall annual turnover of £500 million (nearly £14 billion today). Success of the plan would depend on its ability to increase use of the railway system for both passengers and freight.

In its review of motive power modernisation, the report was unequivocal about the need to replace steam. Whilst cheap in construction costs, it was dirty and pollutive, unattractive to work with, dependent on expensive coal supply and inefficient in operation compared with electric and diesel engine power. Electric power was considered an ideal in every respect, even though the infrastructure involved was expensive. With nuclear power generation around the corner, this was to be the standard for trunk routes and particularly for suburban lines. It is worth noting here that no consideration ever seems to have been given to the use of cheap hydro-electric power, available in Scotland in abundance by that time. In the short term, pending electrification, diesels could more easily replace steam engines so the future would be hybrid.

As far as electrification was concerned, a number of lines were identified for trunk route and suburban construction in the immediate future. These included 190 route miles of Glasgow suburban line based on the Inglis and Stewart Report recommendations but subject to discussions with Glasgow Corporation in respect of the integration of services (presumably through the joint committee reported by *Railway Magazine* in November 1954). The total route mileage was more than the modest first two stages of the Stewart Report proposals.

Technical detail for the electrification elements of the Modernisation Plan was offered in a follow-up report published in 1956, which built upon and seriously amended the BR 1951 electrification report (see Chapter 3).[18] The long-favoured 1500 Volt (V) Direct Current (DC) powered scheme (already in place on the Sheffield–Manchester–Wath line which had opened in 1954)

Left: BR Modernisation map of existing and proposed electrification 1960 (author collection)
Right: A diesel-hauled express train on a 25 kW electrified line from publicity booklet *All Along the Line* 1959 (BR: author collection)

was set aside in favour of a 25 Kilowatt (kW) alternating current (AC) system. This, it was argued in detail based on developments in France, was cheaper in terms of both construction and running, including cheaper locomotives. Rail electrification requires railway-owned substations to prevent voltage drop from the connection point with the National Grid, and the actual collection point between train and overhead wire. With DC systems the distance between the National Grid connection and the railway substation network was 17 miles, whereas with the AC system this could increase to 250 miles, reducing costs significantly for this aspect. However, the higher voltage with AC systems required larger clearances under bridges, through tunnels and other structures, so this added to construction expense. The report was principally concerned with this change of direction and only briefly described the lines proposed for 25 kW AC electrification, including the Glasgow suburban lines. However, an accompanying map showed all the lines recommended by Inglis for stage 1 and 2 electrifications, a map still in circulation as late as 1960.

The End of the Trams and the Rise of Car Ownership

Key to the electrification plans was the aspect of mutual cooperation with Glasgow Corporation, and although there are no reports of this happening publicly as the Modernisation Plan began to be implemented (described in the next chapter), it does seem that Corporation Transport tram services routes beyond the city boundary began to be curtailed from 1956. Closures had started as early as 1949 and there were no renewals of tramcars after 1954. Bringing City buses and trams to within the city boundaries was envisaged by Inglis as part of the plan to bring passengers back to suburban trains, and most such routes were tramways. Glasgow had clung to its trams longer than other British cities, and their demise was regretted by many then and since – a quarter of a million people turned out to watch their final parade on 4 September 1962. Strangely many of those who look back with nostalgia on the days of the Glasgow trams, welcomed the advent of the Blue Trains in 1960/61, which were really the final nail in their coffin, and overlook their direct connection.[19] Had the bold post-war transport planning leading up to and including the Inglis Report been discussed more thoroughly and implemented as recommended, the trams would have been replaced more effectively and comprehensively, and we might not be continuing in the present day to address the same issues around urban congestion. Proper regional integration and planning was not to occur until after 1974 and the creation of Greater Glasgow Passenger Transport Executive (GGPTE – later to become Strathclyde Passenger Transport Executive – SPTE), itself destroyed by bus de-regulation 12 years later in 1986. This though, is again jumping ahead of the story.

The demise of the trams was driven more by the growth of motorcar use than by a push for railway renewal. Eric Fitzpayne's 1956 annual report to the Corporation cited the number of tram accidents involving collision with cars: 5,694 over the preceding year, which he blamed on the proliferation of parked cars on the streets including the main thoroughfares. Many of these resulted in expensive claims by victims against the Corporation – £53,924 over the previous year paid out due to tram-related accidents, amounting to over half the amount paid out for all transport claims. Glasgow, he calculated, also had the worst record for pedestrian casualties from road use.[20]

Argyle Street, Glasgow 1960 (Geograph: Wikimedia Creative Commons)

Fitzpayne announced in 1957 that trams would be eliminated by 1972, but this was soon amended to 1963. The *Glasgow Herald* commented that the Corporation would be held to this promise, the general consensus being that traffic would be speeded up by the removal of the obstacles to movement the trams presented. At the time of the end of services the Corporation produced a booklet celebrating tramway heritage in the city, but even this ended with a belief that trams would never run in Glasgow again, that they had 'reached the terminus'.[21]

Trams are back in vogue, now sitting alongside schemes to reduce car use in city centres elsewhere in Britain, but are expensive and disruptive to construct. Starting from zero (with the exception of the Blackpool tourist trams) systems have been developed in five urban areas of Britain. They are now accompanied by stringent road vehicle traffic management schemes and enjoy dedicated tracks over some of their routeways. They are controversial as seen in Edinburgh, still Scotland's only example – recent schemes for Glasgow have so far never got beyond initial planning stages, a matter we shall return to at the end of this book.

Ex LNER tank engine with Springburn suburban train Glasgow High St. Station 2 March 1959 (© Rail-Online)

There seems little doubt that the increasing popularity of cars added to the drag on proposals for electrification in Glasgow due to the promotion of plans for better road systems. Although in Glasgow car ownership was (and remains) relatively low, it was an aspiration of many; and few, unless suffering directly through dislocation, objected to the road improvements and motorway building that took place between the 1950s and 1990s in the region. Car ownership increased in Scotland from 176,166 to 446,270 ten years later with numbers increasing year by year through the 1960s.[22] Although Bruce's drastic re-construction of Glasgow City Centre never took place, the city was bisected by the M8 motorway, flattening homes and historic buildings in its wake and bringing emissive pollution to many communities. The roots of this lie in both Bruce and Abercrombie; although highways development, including the M8, the Clydeside Expressway and Clyde Tunnel, were not begun until the 1960s, car ownership was being heavily advertised by then and, no longer the preserve of the middle and upper classes, largely replaced motorcycles and bicycles as a means to get to work for those not using public transport (during the 1950s car ownership in Britain doubled). By this time, too, much freight movement was not just by road at the start and end of a journey but was turning towards truck delivery for all types and distances. The peripheral housing schemes were not well served by buses, and this encouraged car

Motorway construction in Glasgow – M8 northern approach to Kingston Bridge 1965
(Geograph: Wikimedia Creative Commons)

ownership, a factor that applied particularly to new towns with their car-friendly road systems. Later, from the 1980s onwards, retail parks began to replace local and city centre shops, their use requiring private car access. The beginnings of mass use of air transport, as cheap foreign holidays became the norm for those in work, required vast areas beside them turned over to car parking: none of the three central Scotland air terminals had rail connectivity (remedied later at Prestwick and more recently in Edinburgh by the new tram system).

In the 1950s and 1960s the negative impact of extensive car ownership was experienced through road congestion and the high risk of death and injury through accident. Concern about the impact of the internal combustion engine on air quality and health was not talked about commonly, and when it was people blamed steam locomotion, coal-fired power and home heating: steam railway locomotion was now being phased out and coal was also seen as in need of regulation and eventual replacement – hopes then being on electric energy production through nuclear power, and later North Sea Gas. Road improvement was seen as a high priority, as was the development of car safety, regulation (such as the introduction of seat belts and laws against drink-driving) and reliability. Cars were becoming increasingly seen as essential to a quality of life, whether or not car ownership was actually an affordable possibility, which for many in Glasgow it was not – a theme returned to at the end of the book.

But Where Were the People?

A central theme of this book is about how the aspirations and efforts of the wartime generation inspired and influenced post-war developments and planning. Chapter 2 discussed how citizen participation in the shape of the post-war world was central to army education, if less explicit for the civilian population. Civilians were asked instead to put their faith in the politicians who would shape the welfare state – a mood successfully captured by Labour with their 1945 landslide election success.

As noted earlier, the plans for nationalisation and their execution did not involve any element of workers' participation despite earlier lobbying by the unions, and government appointed top management, who chose those who worked under them, and so on down to frontline workplace level: a practice continued throughout British Railways days. This of course was the norm in all the nationalised industries, mirroring exactly the way private companies were organised – with Government substituting for shareholders. The hope, though, was that managers were imbued, as some indeed were, with a 'public sector ethos' (a term rarely heard nowadays). This method of running public services has always been criticised by those on the left, but no Labour government to date has sought to venture down the road of workers' control.

What though of popular involvement in planning by the people who would use services? The Bruce and Abercrombie Reports were mainly undertaken in wartime so their committee members might be forgiven for not asking people of their views as part of the process. The 1948 Fitzpayne Report was undertaken by a public servant (of Glasgow Corporation) reporting to a public committee of elected councillors. They would no doubt have argued at the time that they represented the voice of the people.

By the time we get to Inglis in 1951, public consultation was as it always had been with government organised committees – notably absent. Expert advice might be sought, and indeed was, but not the view of the consumer. There is no evidence amongst all the information brought together by Inglis that public opinion was surveyed systematically or otherwise. Instead, examination was made of traffic figures and ticket sales, and trends calculated from these. Certainly, the evidence that improvements elsewhere, including overseas, had brought people back to railway use, formed part of the Inglis argument for change. The people of Glasgow, though, were never consulted. The last of the reports considered, the Stewart Report of 1954, was a shorter feasibility study, but again included no foundation in public opinion.

The 1947 Transport Act did introduce Transport Users Consultative Committees (TUCCs) on a central and regional basis (i.e. one for Scotland) which were intended to bring the voice of consumers into the industry. These included numerous transport officials and representatives of interest groups including users from agriculture, industry, the trade unions and local authorities, who served for a three-year period. These were broadened in terms of powers in the early 1960s in the wake of the Beeching report on railway closures (to offer views to the Minister of Transport on the impact of rail freight closures as well as passenger services). There is no evidence through the reports cited that they brought any bearing on any of the work undertaken by the bodies involved and in Parliament in 1950, were described as 'farcical' during a committee discussion, and the 'local' (i.e. regional) committees were only set up that year.[24] In 1958 they were criticised in Parliament again for absence of the voice of railway passengers, and in the words of Fenner Brockway Labour MP, for the fact that '...not one passenger in a thousand... has any knowledge of these committees.'[24]

This is a matter affecting not just transport challenges and solutions in the post-war period, but all other aspects of planning too. The Bruce and Abercrombie plans all involved mass population displacement and rehousing. Those involved were never asked for their opinions but left to cast votes at council and Parliamentary elections for those who would decide for them – even though the details of such plans were never included in party manifestos or individual pledges. In the mix too were the conflicts and interest-driven concerns of different layers of the bureaucracy: Scotland against England, the Labour Glasgow Corporation against the Tory governments of the 1950s and early 1960s, and Glasgow versus Edinburgh.[25] In fact, systematic public consultation, which we take for granted today (even though we might not trust the means and methods), was not to come about until the Housing (Scotland) Act of 1974 which established Housing Action Areas required this as of right.[26]

The city officials, planners, politicians at all levels and others responsible for the way Glasgow met with the challenges of the post-war period made many mistakes, which still affect lives today. The development of suburban

The new Glasgow: Red Road Flats under construction 1964
(Miles Glendinning: Wikimedia Creative Commons)

railways through the Blue Trains initiative that followed the BR Modernisation Plan was an example where they got it right, even in modest terms and in the absence of real consultation. This is the subject of the next chapter.

Chapter 5

BLUE TRAIN DAWN

'Out of a cutting above the River Clyde and into the sunshine there flashes a train at speed. Its coaches are painted the bright blue of the kingfisher, gay against the softer browns and greens of the fields, and its windows, almost unbroken in length, flash cheerfully. An old man in a chair at a cottage door on the hillside looks down to see that the three coaches ripping along so fast and so purposefully do so as if by dark magic; vaguely he misses the smoke and steam, the glowing fires and heaped coals of the engines of his young manhood. At least he has lived long enough to see the electric train come to Scotland.'[1]

Introduction

A key concern of the 1955 Modernisation Plan, as discussed in the last chapter, was the pace in which steam might be replaced with electric and diesel motive power. Initially this moved in a manner aimed at thorough evaluation of various designs and modes (e.g. diesel electric and diesel hydraulic). However, mounting financial problems determined a belief that passengers and freight users could only be attracted back to railways by accelerated modernisation. In 1958 the decision was quietly taken at the highest level that resulted in the abandonment of caution, and the rapid introduction of diesel locomotives that had not been tested in practice. Whilst this pre-empted the end of the steam locomotive, including the early scrapping of types that were new and working efficiently, it wasted resources on numerous diesel types that were

A sunny but smoky scene in Glasgow Charing Cross Station 2 July 1958 (© Transport Treasury: WAC Smith)

taken out of service early in their lives. Tory government policy towards the railways from 1951 onwards, whilst not (at that stage) questioning nationalisation, was markedly different to that of the Labour administration who had introduced public ownership. There was a move to decentralise control and allow each region to determine what it needed – with poor results in some cases when it came to decisions about motive power.[2]

This had some negative consequences particularly in the Western Region, but also in Scotland: the adoption of ultimately rejected diesel-hydraulic models furthered the demise of the once mighty North British Locomotive Company in Glasgow. They had invested in this mode when their foreign steam engine orders went into terminal decline through the 1950s. It also resulted in wasted expenditure and the sight, in some cases, of diesel locos being towed to the scrapyard by the steam engines they were designed to replace. Mainline electric introduction was focused in the late 1950s, and into the 1960s, on the London Midland Region and the West Coast Routes, initially in the Crewe, Liverpool and Manchester areas. Here, there was a smoother transition.

All clouds have silver linings and whilst the results of diesel introduction were not spectacular (including in Scotland), the delegated authority turned out well for the Glasgow suburban trains. This chapter describes the design, the construction of the infrastructure required for electrification, and the introduction of the Blue Trains. Whilst there was still an absence of the transport co-ordination envisaged by the post-war reports, this development demonstrated that railways could have a vibrant future. The words at the start of this chapter, from the British Railways' introductory publicity book for the new trains, might not be Shakespeare or Burns, but they do capture some of the excitement of entering a new age associated with Glasgow's first advertised electric services in November 1960.

The Go-Ahead for Glasgow

The Modernisation Plan launched several electrification schemes, only one being in Scotland; the others were in the Eastern and London Midland Region. This last featured main lines linking Crewe and Manchester in the north-west of England which required the development of mainline electric locomotives – a project that was to garner attention and become the showcase for mainline electrification. The cost of the North Clyde project, which would be constructed

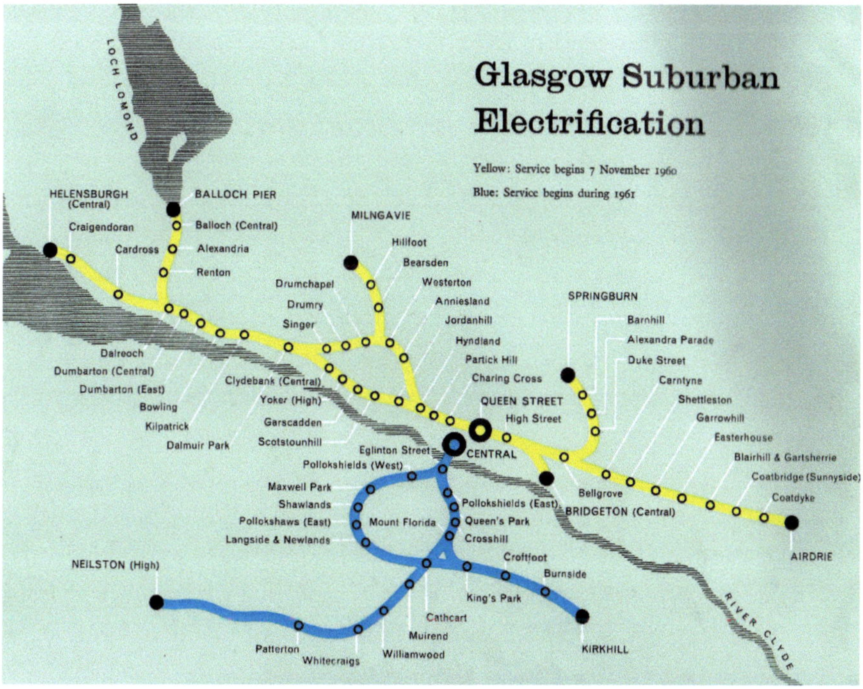

1960 Glasgow electrified lines (BR *Glasgow Electric:* author collection)

initially as Stage 1 of the Scottish scheme, was determined as £16 million (£367 million today) – an increase from the £11 million estimate offered in 1954.

This was to be a network of electrified lines serving a working-class urban population who were expected to use it to get to work – six electrified lines were run into a terminus just off the main line at the Singer Works in Clydebank, which would involve timetabled works trains. Other custom was expected from middle-class commuters on the Milngavie branch, and also from Helensburgh. The latter, and the Balloch branch, would also attract Clyde steamer and holiday users. This combination assured success.

Tensions between Glasgow Corporation and BR (discussed in the last chapter) over transport planning for the city were, the public were reassured in press reports in 1956[3], not delaying plans for electrification, and agreement was reached with Glasgow Corporation and construction work commenced in 1957. Stage 1 would involve the North Clyde service through Glasgow Queen Street Low Level from Airdrie to Helensburgh with the branches to Springburn, Milngavie and Balloch, and the Southside service from Glasgow Central around the Cathcart Circle and out to Kirkhill and Neilston High, with services to start in a hoped-for four years. It was reported at that time that plans were under development for Stage 2 involving the other lines proposed in the Inglis Report: from Glasgow Central and St. Enoch out to Wemyss Bay, Gourock, East Kilbride, Kilmacolm, Kilmarnock, Hamilton and Motherwell.[4]

By 1957 the financial allocation and the amounts committed for the other schemes were in danger from the BTC (and behind them the Government) due to renewed and continued problems caused by reducing revenue across BR, mainly a consequence of rapidly reducing freight traffic. However, the Scottish Region were successful in arguing that the consequences of cancellation would reduce revenues even further. In the next few years, similar consideration was given to suspension, but by 1959 orders had been made for trains and equipment, and construction was well underway, so Stage 1 was allowed to proceed. Publicity and promotion did not get into gear until 1960, but there was a degree of public awareness of the modernisation plans, and cancellation would have produced an adverse press reaction and shaken confidence in the railway system even further. In fact, there was a push to complete on time and introduce services in 1960, and the same applied to other schemes, with the consequence, as we shall see, that some of the equipment went into service inadequately tested.

The Work of the BR Design Panel

A very successful spin-off from the Modernisation Plan was the establishment in 1956 of the BR Design Panel, arising from the desire of the British Transport Commission that the new motive power replacing steam should be modern in design and of a standard that would be pleasing and of a quality that would stand the test of time. This would engage with criticism that British Railways' design did not match progress and looked generally dated. Indeed, the pragmatic prediction of railway enthusiasts at the time of nationalisation in 1947 was that '... the dull uniformity of standardisation will be offset by greater efficiency in operation.'[5] Although initially concerned with train design, the team soon ventured into all aspects of the way the railway was seen by the public, from logos and signage to uniforms and sundry equipment. By the 1990s and privatisation, the work of the Design Panel was renowned across the world as an example followed by other rail operators.[6]

An early beneficiary of the team's work was the design of the new trains for the Glasgow electrification, and the promotion that would help sell them to the public. The design and shape of the new trains did not happen by accident as will be seen below – others that emerged in the same era for newly electrified lines were comparatively bland and ordinary.

Eastern Region 1959 EMU – utilitarianism avoided for the Glasgow design (© ColorRail)

To ensure that the laid down specification was met, for the very first time in Britain full-size mock-ups of the carriages were constructed which were inspected by railway officials and others involved. This included ASLEF officials who apparently approved of the see-through drivers' compartment and wrap-around windows (later removed as described below). The doors, unlike those on the long-established Southern Region lines, were sliding and operated by the guard who had a separate compartment in the centre driving trailer of each three-car set, over which the overhead pantograph was located – lowering the roof line in that section only. There was no corridor, and each passenger compartment was self-contained, spacious and bright with glass partitions around the doors – two on each side of the front and rear carriages. The colour of the train sets was a matter of discussion and consideration and eventually a version of Caledonian Railway Prussian Blue was decided upon. The advice was sought of C. Hamilton Ellis, the Scottish-born railway historian, and he advised that there were various versions of the CR, depending how much white lead was added to the mix at the works at St Rollox in Glasgow. The Design Panel then concocted their own, which they called 'Electric Blue'.[7] The new system was, it was decided, worthy of a new corporate image for the stations including a logo and new lettering style for the station signs – all produced in glass fibre rather than traditional metal.

The names of those responsible for the various aspects of the design should be recorded:

> J.F. Harrison, BR chief mechanical engineer: overall direction;
>
> S.B. Warder, BR chief electrical engineer in conjunction with Metropolitan Vickers: electrical design;
>
> Wilkes and Ashmore, consultant designers: the exterior design of the train sets, the front-end detail, the driver's cab and the passenger seating;
>
> Misha Black, DRU and John Diamond of Diamond, Hodgkinson and Partners: the interior detailing and finishes;
>
> F.H.K. Henrion: the station signs and logo, and gave advice on the colour scheme for the trains;
>
> George Williams of the BR Design Panel provided overall liaison.

The designs that were agreed, little apparently changing because of the consultations, were ready for construction by 1959. The electrical equipment work was undertaken by AEI Metropolitan Vickers (Metrovick) in Manchester, who had equipped the 1500 Volt DC locomotives built at BR Gorton Works for the Manchester–Sheffield–Wath line. In a gesture that gave some ownership and investment in the project to Scotland, the 91 Electric Multiple Units (EMUs) were built at Pressed Steel in Linwood (all the others for the London Midland and Eastern Regions were built in BR's own works in England). The Linwood plant was established in 1937 by the Glasgow shipbuilding company William Beardmore for defence work but was transferred to Pressed Steel ownership by the Board of Trade in 1947 to address unemployment in the area. Initially constructing railway rolling

November 1959 *Design Magazine* cover (author collection)

stock, it began building truck and car bodies for various manufacturers when railway work declined, until taken over by Rootes-Chrysler in 1967, eventually closing as a car plant in 1981.

The result was ahead of its time and quite unique – if matched in its day only by the Pullman trains – diesel multiple units (also blue) designed for mainline express services from London to the north and West Country in England, and described as 'luxury'; these began operating in July 1960 and lasted until 1973. In contrast the Blue Trains in Glasgow were in service for over forty years. They still look modern and impressive, a tribute to their designers.

The 1960 Blue Trains were designated by BR as Class AM3 EMUs (later changed to class 303, the name popularly used today) and there can be no greater contrast in terms of design flair than with the Class AM2 (later named 302) Electric Multiple Units built for the Eastern Region services that were introduced into service a year earlier in 1959. These were blandly traditional (i.e. like the ones brought into use in the 1950s for the extensive Southern Region services), and unattractive in their standard green paintwork. The Blue Train designs were largely copied for other electrifications on the London Midland Region routes later in the 1960s, and traces of their design concept can be seen in electric train units built for suburban services throughout Britain in subsequent years and up to our own times.

James Ness

A perhaps overlooked aspect of the Blue Trains' development was the part played by BR Scottish Region's general manager James Ness, who was appointed in 1955 on the sudden death of his recently appointed predecessor, A.E.H. Brown. Ness, at 52 years, was young for such a role in that era. He had started his career with NBR, and so was steeped in the steam tradition. However, he recognised opportunities for marketing and popularisation that went unheeded by colleagues in other regions. One example was his order for the restoration, maintenance and use of four previously preserved steam locomotives from the main pre-1923 Scottish companies for special use and rail tours. This was a very successful initiative with Ness actively encouraging their booking by railway enthusiast societies; the sight of these highly polished and colourful engines won positive public interest in the railways, but was not replicated to the same extent elsewhere in Britain.

Top: Mock-up of the Glasgow EMU design (*Design* Nov. 1959: author collection)

Bottom left: Seating (*Design* Nov. 1959: author collection)

Bottom right: Construction underway at Pressed Steel Linwood (BR *Glasgow Electric*: author collection)

On similar lines Ness is credited with providing early assistance to the nascent preservation movement in Scotland – identifying the Falkirk location which the Scottish Rail Preservation Society (SRPS) used as storage and a depot for many years until its permanent site in Bo'ness was developed in the 1980s.

With regard to the Blue Trains story, Ness is credited with responsibility for the brief to the Design Panel. He insisted that the Glasgow trains should be not only of advanced technical design, but should offer 'high standards of comfort, convenience and pleasure in travelling.'[8] His specification included taking advantage of the beautiful countryside on the Clyde outside the city by offering good front and rear views from the carriages – in effect observation cars. Advantage in this respect was taken of the fact that the existing lines in the Glasgow area were three inches wider than the general width for BR trackwork, so the width of the new trains was to be nine feet three inches instead of nine feet.

When the forced withdrawal of the electric services took place in December 1960 (see below for more detail), Ness took the decision, after rapid consultations with staff and the BTC, to suspend the service in the early hours of a Sunday morning, and successfully roused and inspired his staff to reintroduce a steam service by the following Monday morning, involving many, including no doubt himself, in continuous work for 24 hours and more.

Despite his notable achievements and the respect of many staff, Ness was not universally liked: he was described by a colleague he managed as '...a complete autocrat but a man of constant drive and immense ability.'[9] Gourvish, whose account of the BR years is held in general regard by rail historians, described Ness quite unfavourably without acknowledging any of his achievements in Scotland.[10] In 1963 Ness was promoted to a post with the BRB as board member for planning under Dr Richard Beeching (see next chapter). He retired early on Beeching's departure in early 1965, with a reputation apparently tarnished for many by his association with his unpopular chairman.

Building the New Railway

Construction on the lines commenced in 1957, using methods and means developed in previous electrification schemes. In order to avoid having to increase the height of the tunnels and bridges any more than necessary in the city centre and short branches, voltage there was reduced from 25 kV

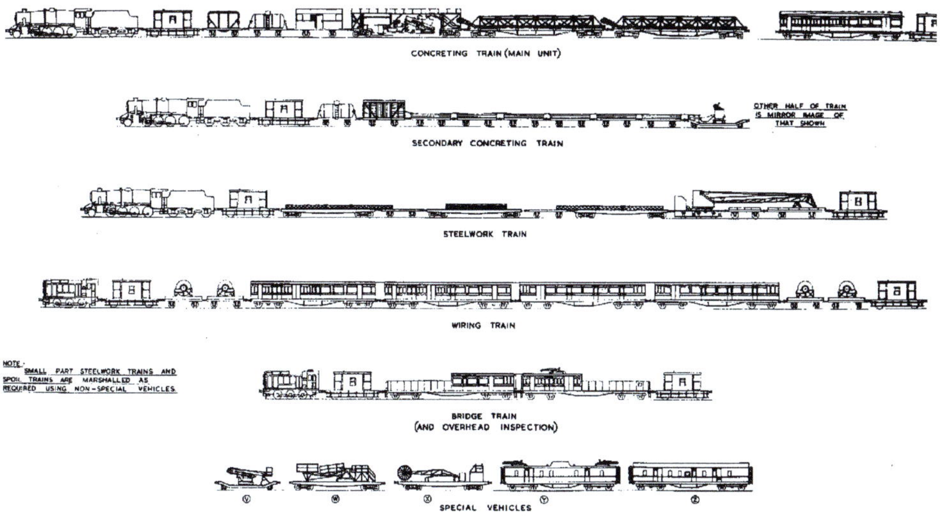

Electrification construction train types (BR *Papers of the Electrification Conference 1960*: author collection)

to 6.25 kV. Trials had shown that with a high voltage system 11 inches of clearance was necessary above the electrified lines in order to avoid flashover (with damaging consequences to the wires), but a reduction in voltage eliminated this problem and saved expenditure where there were extensive tunnels and bridges. This was a standard applied elsewhere in urban areas until technical development following the transformer problems (partly arising from the switchovers), described below, rendered this unnecessary several years later.

The 20-minute British Transport Film *Report on Modernisation – 1959* with its monotone dialogue spoken against a background of church organ music, includes a section celebrating electrification. It contains a short clip of one of the new Glasgow train sets and, separately, construction of electric overhead wires filmed in the Crewe area of north-west England. This is alarming by today's standards of health and safety regulation: men in shirt sleeves are seen standing in a line on the roof of a converted carriage threading the overhead wires from a large reel through gloveless hands, lacking helmets, guard rails and any other safety equipment; the commentator remarks on the hazards involved, but does so flippantly in misplaced admiration of the physical prowess involved. These workers were employed by private contractors, but standards for BR staff were little better. They may have been paid well by

LEFT: Glasgow north track lowering (BR *Glasgow Electric*: author collection)
RIGHT: Wiring train work underway (BR *Papers of the Electrification Conference 1960*: author collection)

the standards of the day, but their manual labour methods seemed little changed from those of the navvies of the mid-19th century.

The reconstruction of the line from Airdrie to Helensburgh involved a great deal of work. Although this is not intended to be a technical book, it is worth listing some of this to illustrate the extent and magnitude of the human effort involved. The installation of electric signalling involved the reduction in signal boxes from 56 to 27 and the construction of two new large power (rather than mechanical) signal centres at Hyndland and Dumbarton, and other new signal boxes in Airdrie, Bowling and Helensburgh. A new maintenance depot for the trains was built at Hyndland along with a new through station to replace the existing terminus. A new junction at Bowling was installed so the line would run through Dumbarton East Station (joining together lines from the rival CR and NBR companies). Lines were lowered under bridges and through tunnels in 28 locations, and bridges were reconstructed or removed in 119. The installation of overhead lines and the posts that would carry them was a major and disruptive exercise. Some stations were redesigned to give the route a clean and modern look with the new logos. The electrical work involved feed-in from the National Grid and control of the current: a control centre was built in Cathcart (for North Services as well as projected South ones) and duplicate feeder stations in Dalreoch (Dumbarton) and Parkhead, and sub-feeder stations in Airdrie and Westerton. This would ensure that the supply could be maintained throughout the system in the event of failure in any one location. All this took three years but was completed on schedule; perhaps, as already mentioned, it was all too quick: the technology was new to the engineers who built the motive power equipment based on French experimental designs, and it was new to the railway staff as well. Meanwhile similar construction work had started on the Southside route.

Trials and Introduction

Scottish Region management were rightly proud when they were able to show the new trains to the public in Glasgow for the first time in September 1959. The display of a new electric three-coach trainset along with other up-to-date motive power and equipment in two set-aside platforms in Central Station was associated with the Scottish Industries Exhibition in the Kelvin Halls. The modern look given to this exhibition was in marked contrast to the 1951 Festival of Britain effort. In order to maximise attention on the new train and the other exhibits, BR ran a special shuttle service to Kelvin Hall Station

Left: The 1959 Glasgow Central Station Exhibition (BR Scottish Region staff magazine Nov. 1959: Stuart Mackay collection)

Right: Sir Brian Robertson and senior BR managers at opening of the 1959 Central Station Exhibition (BR Scottish Region staff magazine Nov. 1959: Stuart Mackay collection)

from Central Station. Sir Brian Robertson, the Chairman of the BTC, spoke publicly at the opening of the exhibition, with a clear appeal to Glasgow's people and their political leaders:

'This great project had its origins in discussions I held with the authorities in this city four years ago. I pointed out then that this would be a very expensive project and could only be justified on the assumption that the services, once provided, would be fully used. Based upon the assurances of co-operation I received at that time, the Commission felt able to proceed with the work, and its realisation is now evident. I can say that there will be no better rolling stock on any suburban service in the kingdom, and I doubt that there is any as good anywhere else – and it has been built in Scotland. Services on this first stage will start next year... The further extension of this electrification will depend on the encouragement we receive from the people of Glasgow and surrounding districts. My own view is that by making use of a modernised suburban electric network, the City of Glasgow could do more to solve its traffic problems than by any other single means. We will ensure that the service given is good, and it will be up to the people of this city to make their own choice.'[11]

In 1959 the Scottish Region's plan for motive power replacement suggested that the enormous deficit accrued through modernisation would be over by 1963, when passenger service income and expenditure would balance out, so the importance of success with the electrification in the Glasgow area would have been crucial to ongoing financial viability at a national level.[12]

By 1960 the work on the lines had reached the stage when the train sets started appearing on trial and staff induction runs, allowing publicity photo shots that appeared in a 40-page glossy brochure that was published in November: *Glasgow Electric*.[13] The cover featured a truly inspiring painting by railway artist Terence Cuneo that was also used for BR promotional posters. The press started covering the soon-to-be new service, one aspect that drew comment being the abolition of first-class accommodation in the new train-sets. Whilst most people recognised that the comfort and style now on offer rendered the new arrangements a vast improvement on steam-age

OPPOSITE: Clockwise from top left: Hyndland Depot 1961: Rodger Whip/Darrel Hendrie collection, 1960 training run Cardross: © Transport Treasury/WAC Smith, BR postcard 1960: author collection, Balloch 1960: © Rail-Online, 30318 refurbished 1991: Darrel Hendrie, Craigendoran Pier c1961: Rodger Whip/Darrel Hendrie collection, Bridgeton Central 1960: © Rail-Online, ceremonial opening train 5/11/60: © Transport Treasury/ WAC Smith

26

A1

A1

14

26

BALLOCH

GOUROCK
50

carriages, there was an adverse reaction from others. A letter published in the *Helensburgh Advertiser* on 21 July 1960 went as follows:

> *Sir,*
>
> *Regarding the furore of classes on the local line: I am sure – no positive – that the second class peasants will object to travelling in this classless carriage. The people, the proletariat, have a horror of travelling with their betters, any social convenience or society which they think is beyond their dim reasoning! The dignity of the so-called working man is a myth which draws the wavering line at meeting his betters on an even social scale. I am in the nominal position of offering to BR a certain amount of hard cash to convey my person to Glasgow in greater comfort than the ordinary masses. Therefore, why should I have to subject myself to the riff-raff that screams rants and swears in its second class crummy carriages.*
>
> *I want no part of the masses and, considering their outlook on life, they want no part of me. So, for hard cash and not principles of classlessness, let's have the first-class carriages back and let the common herd travel in their cattle trucks.*
>
> *Yours*
>
> *'Bowler-hatted Brigadier'*

Commencement of Electric Services

The new timetable for the Airdrie–Helensburgh line was published in the autumn of 1960, with a start to electrified services from Monday 7 November (in fact they started on a trial basis two days earlier). The preceding timetable based on steam motive power took 1 hour 27 minutes for a through journey with no changes, offering this only irregularly throughout the day on weekdays. The new one offered this service twice-hourly with a journey time of 1 hour 17 minutes – a vast improvement all round. The new timetable outlined similar improvements for the branches to Balloch, Milngavie and Bridgeton Central – the services all operating the same route through Glasgow Queen Street Low Level between the new station in Hyndland and High Street.

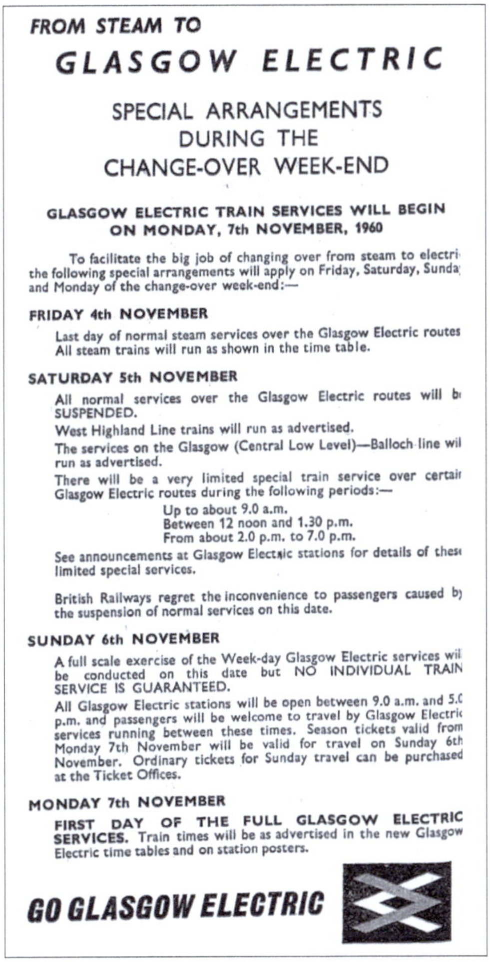

FROM STEAM TO

GLASGOW ELECTRIC

SPECIAL ARRANGEMENTS
DURING THE
CHANGE-OVER WEEK-END

GLASGOW ELECTRIC TRAIN SERVICES WILL BEGIN ON MONDAY, 7th NOVEMBER, 1960

To facilitate the big job of changing over from steam to electri the following special arrangements will apply on Friday, Saturday, Sunda and Monday of the change-over week-end:—

FRIDAY 4th NOVEMBER

Last day of normal steam services over the Glasgow Electric routes All steam trains will run as shown in the time table.

SATURDAY 5th NOVEMBER

All normal services over the Glasgow Electric routes will b SUSPENDED.

West Highland Line trains will run as advertised.

The services on the Glasgow (Central Low Level)—Balloch line wil run as advertised.

There will be a very limited special train service over certai Glasgow Electric routes during the following periods:—

Up to about 9.0 a.m.
Between 12 noon and 1.30 p.m.
From about 2.0 p.m. to 7.0 p.m.

See announcements at Glasgow Electric stations for details of thes limited special services.

British Railways regret the inconvenience to passengers caused b the suspension of normal services on this date.

SUNDAY 6th NOVEMBER

A full scale exercise of the Week-day Glasgow Electric services wil be conducted on this date but NO INDIVIDUAL TRAIN SERVICE IS GUARANTEED.

All Glasgow Electric stations will be open between 9.0 a.m. and 5.0 p.m. and passengers will be welcome to travel by Glasgow Electri services running between these times. Season tickets valid from Monday 7th November will be valid for travel on Sunday 6th November. Ordinary tickets for Sunday travel can be purchased at the Ticket Offices.

MONDAY 7th NOVEMBER

FIRST DAY OF THE FULL GLASGOW ELECTRIC SERVICES. Train times will be as advertised in the new Glasgow Electric time tables and on station posters.

GO GLASGOW ELECTRIC

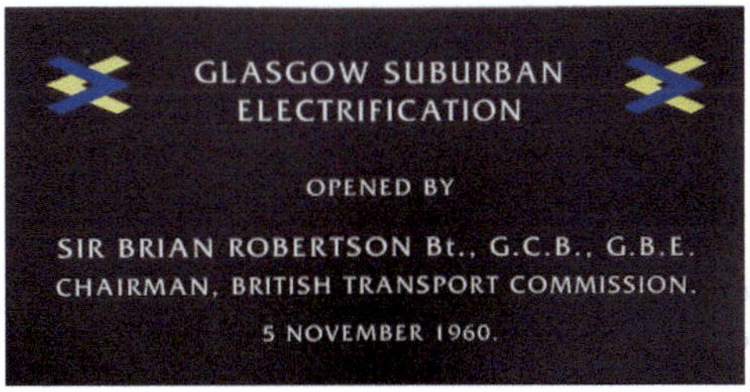

LEFT: Newspaper advert advising of opening arrangements November 1960 (Darrel Hendrie collection)

TOP RIGHT: Official opening at Glasgow Queen Street 5 November 1960 (BR Scottish Region staff magazine Nov.1960: Paul Strathdee collection)

BOTTOM RIGHT: Blue Train opening plaque in Queen St. Station (author)

Services from Queen Street were described as 'Eastbound' and 'Westbound' and mapped as such on large hand-painted murals on the two low-level platforms (covered by yellow panels in the 1970s). Trains from Helensburgh to Glasgow increased in frequency from 24 to 37 on weekdays, the journey being now 13 minutes faster.

A ceremony involving the unveiling of a plaque by BTC Chairman Sir Brian Robertson took place at Queen Street Station on Saturday 5 November, and was prominently reported in the Scottish press. *The Scotsman* headline for Monday 7 November read 'Streets for movement, not as garages', reporting that Robertson had called for co-ordination of services which might require the subordination of sectional interests. Lord Provost Dame Jean Roberts (Labour) of Glasgow spoke with similar sentiments and stated that negotiations

Hyndland Depot April 1960 (Rodger Whip: Darrel Hendrie collection)

had involved one guiding principle: the provision of satisfactory services at reasonable cost to the travelling public. After the ceremony, the party, including other Provosts and Members of Parliament, then travelled by special Blue Train to the new Hyndland Depot, stopping at Partick Hill where the rapid braking and quick acceleration were demonstrated.

The Glasgow *Evening Times* newspaper described 'ecstatic praise' amongst passengers it spoke to at Queen Street for its edition on the first day: two male regular commuters for over 25 years said it was worth the wait and were sure the public would support the scheme. A woman passenger, Mrs Mary McPhail of Coatbridge said 'Oh, it is lovely – so comfortable and clean.' However, there were ominous signs of what was to come when a morning train broke down in Westerton, causing the cancellation of the 9.26 to Balloch, and 9.44 to Helensburgh.[14] The article interestingly referred to the new units as 'Blue Trains' – a name that had apparently been adopted by the public after the public exhibition in Central Station the previous year. That name was to stick in association with the original train sets even though a few years later a different shade of blue was adopted by British Rail as standard. The Blue Trains themselves went orange in the 1980s under the corporate identity adopted by Glasgow Passenger Transport Executive (GGPTE).

Meanwhile, in Helensburgh the local newspaper, *The Helensburgh Advertiser*, reported the last steam service on 5 September drawn by loco no. 67655 with driver John Armstrong and fireman James Lardy. The first electric train was in the hands of Willie Tran (even giving the reader his home address in the town!) and the guard was Charles Taylor. Passengers on that train included the Lady Provost of Helensburgh Miss Young, the County Convenor Dr McGregor, Sir Robert Inglis, and commuter for the previous 54 years, Mr Harry Morton.

In the first week of service 400,000 passengers used the new service, in contrast to the 170,000 users in the same period the year before. Off-peak use suggested a huge increase: numbers entering or leaving Queen Street Low Level on one Saturday numbered 40,000, compared to an average of 9,000

TOP LEFT: First and short-lived North Clyde timetable 1960 (author collection)

TOP RIGHT: View through the driver's cab en route to Helensburgh (*Glasgow Electric*: author collection)

LEFT: BR Summer 1963 leaflets advertising sailing connections via Craigendoran (author collection)

with steam services; 50,000 people were now using the Sunday services. The fare system offered was also attractive and competitive. Clearly the people of Glasgow and the communities served were taking to the Blue Trains, and BR staff were said to share their enthusiasm!

From the outset the Blue Trains featured an innovative system of head codes that along with the destination signing above the driver's cab would allow passengers to identify the route of their train as it entered their station and these were issued in a little leaflet at stations (a practice in place on the Southern Railway suburban electric routes in the 1920s). The first digit advised of the destination, and the second the route and stopping arrangements e.g. 15 indicated a Helensburgh bound train, stopping at all stations by way of Singer. This was later altered slightly for the Southside routes, but eventually abandoned altogether given the likelihood that it was understood by few passengers.

Problems!

Issues with electrical equipment occurred within the first few days – what might be expected with a new system. On the first day some blown fuses caused a cancellation as reported in the *Evening Times* article on 7 November, and on 2 December a train brought down overhead lines in Bridgeton. British Railways though were confident that these teething problems could be overcome. However, catastrophe was looming.

The equipment testing prior to the service's introduction had thrown up issues concerning the rectifier and transformer equipment located in the central carriage of the three car sets, where the current from the overhead wires fed through the pantograph over the guard's compartment. Much of this equipment was located underneath the passenger section in this carriage.

Unknown to the public at the time, on 30 October, on a trial timetable run prior to service commencement, an explosion had occurred on an EMU at Charing Cross involving the transformer equipment (this alarming event was made public at the public inquiry and reported in the *Evening Times* on 21 December 1960). This was put down to poor venting and adjustments were made to all units and the problem was believed solved. A less dramatic incident occurred on 14 November when a train approaching Queen Street station suffered a loss of power; no serious damage occurred but again the fault lay with the transformer equipment.

On 13 December a serious accident occurred at Renton on a six-coach train (two EMUs coupled together as was practice on busy services) carrying 230 passengers: an explosion occurred involving the same equipment that seriously injured the guard and two passengers. The driver, William Clark, told the public inquiry (reported in the *Evening Times* on 22 December 1960) that the train suddenly braked '... as if some hoodlums were having a go at the communication chord.' He ran back having failed to raise the guard on the phone, to find him 'stammering' at the window. Amidst panic amongst the passengers, he placed detonators on the line to protect his train before contacting the signal box in Alexandria. Again, preliminary investigation resulted in instructions to further improve venting. The following day, 14 December, smoke was seen coming from the transformer area on another six-coach train entering Carntyne Station, and it was taken out of service before any further issues developed.

Concern was mounting and it only took one more incident before the electric service was suspended indefinitely. This occurred three days later on Saturday 17 December at Garrowhill, when the 12.45pm six-car train from Helensburgh to Airdrie, with 250 passengers, was seen to be emitting smoke from the same location. The fire was extinguished quickly by the driver, alerted via direct telephone by the guard, and no one was injured. In the early hours of the following morning the Scottish Regional Manager convened a meeting of his departmental managers, and a decision was made at 01:38 to withdraw the electric services pending full investigation and whatever remedial work was necessary. The incidents listed involved five different units, and by now a total of 17 were unserviceable, so strain on the service was being experienced anyway.

In an amazing effort by staff at all levels, a reduced service was put into place on the Monday morning – the new timetable being made public within hours of the decision to suspend the electric services. This was remarkable considering that the introduction of the electric service had seen the redeployment of 37 steam locomotives, the closure of the facilities that serviced them at Airdrie, Helensburgh and Balloch, and the redeployment of 100 staff. The system had to be cleared of all the electric units, steam engines had to be diverted from other duties elsewhere, and carriages sought, prepared and heated.

On the afternoon of Sunday 18 December, the BTC convened a meeting of all the manufacturers, and it was decided to examine all the damaged transformers at the AEI Wythenshawe transformer works in Manchester. In a message to staff on 18 December general manager James Ness wrote:

'All Scottish Railwaymen will have learnt with bitter disappointment of the decision to withdraw temporarily the Glasgow Electric Services.

This has been done to enable the manufacturers – AEI Limited – to remedy certain defects.

We have done our utmost to operate this service efficiently and the response of the public has been magnificent. Glasgow Electric raised our morale and showed what the modern railway can do. Now we must in the interests of passengers and staff temporarily lower our flag.

I am proud of the effort made by all staff employed on the Glasgow electric service. Thank you for trying to keep it going. It will run again as soon as the manufacturers have put the equipment right. Meantime let us unite to give the best possible alternative service.

James Ness,
General Manager'[15]

The *Railway Magazine* reporting this story in its February 1961 issue felt that with 100,000 miles of testing prior to introduction, with no serious incidents involving the transformers, no blame could be attached to BR for bringing in the service too quickly. However, this remained the view of some. Investigations and solutions involving equipment replacement on all 91 units were to resolve the issue for all time and in later years the 6.25 kW sections were replaced with 25 kW equipment. The first new transformer was tested and ready on 22 February 1961 and subsequent trials overnight between 21 and 23 March were successful. Seventy-seven units were then rebuilt with new transformers; other modified electrical equipment was installed in each unit over the next five months at the London Midland Region works at Dukinfield, Manchester, and the remainder at Pressed Steel, Linwood. Trial runs with the modified units were exhaustive with 380,000 miles being clocked up by 1 October 1961 when the North Clyde electric services recommenced. The week prior to the service restarting a full timetable trial was run over the North Clyde lines and this passed off without incident.

BR published an internal case study of the experience of the suspension of electric services with comments on how replacement by steam had been

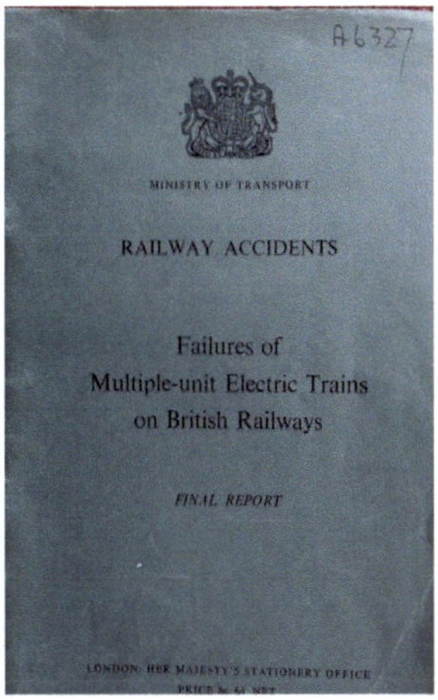

MINISTRY OF TRANSPORT

RAILWAY ACCIDENTS

Failures of
Multiple-unit Electric Trains
on British Railways

FINAL REPORT

LONDON: HER MAJESTY'S STATIONERY OFFICE

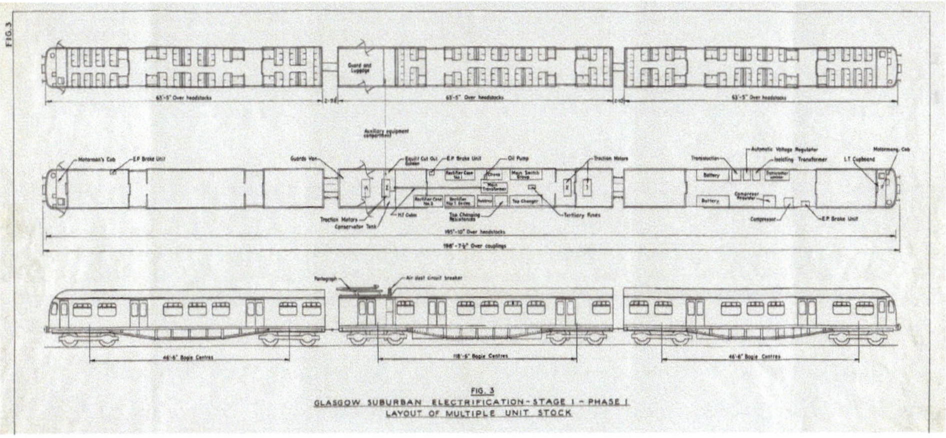

BRITISH RAILWAYS

SCOTTISH REGION

TEMPORARY

ALTERATIONS TO

PASSENGER SERVICES

SCOTLAND

COMMENCING
2nd JANUARY 1961
Until Further Notice

GLASGOW
AND
SOUTH WEST
DIVISION

This Time Table replaces the Glasgow Electric Time Table: also Tables 16, 19 and 34 in the Scottish Region Time Table and Supplement dated 2nd January 1961 due to the temporary withdrawal of the Electric Trains from the Airdrie - Springburn - Bridgeton - Milngavie - Balloch and Helensburgh routes where Steam Hauled Trains will operate until further notice.

TOP LEFT: Blue Train EMUs on tow to Manchester February 1961 after the suspension of electric services (© Transport Treasury: WAC Smith)

TOP RIGHT: The report into the 1960 Blue Train failures March 1962 (author collection)

ABOVE: Detail plan of the Blue Train units showing location of electrical equipment from the March 1962 report (author collection)

LEFT: Hastily prepared BR replacement timetable January 1961 (author collection)

achieved so seamlessly in late December 1960. This concluded that the belief amongst staff that the electric service was a worthwhile aim in the long term enabled massive short-term effort to keep services running. The report noted that '...morale is often most high when physical conditions are most difficult' – an obvious wartime allusion.[16]

The *Glasgow Herald* of 26 September, reporting the reopening of the service the following week, told that heavy materials simulating passenger weight would have to be removed from the carriages, which would require some refurbishment, and the overhead lines and insulators cleaned of the smoke dust from the steam service.[17]

The Electric Service Revived

The *Evening Times* newspaper of Saturday 30 September 1961 welcomed the expected resumption of services the following day with a poem:

> 'Blue Heaven
> Tomorrow I'll blue a copper or two
> On a trip in a train of a similar hue,
> With never a choo choo, never a puff
> As the new transformer does its stuff
> Oh! The luxurious, fast but not furious,
> Speedy and comfy, completely non-stoorieous!
> Let others go wheezily, gear change uneasily,
> Riding in trains, black-smokily, dieselly;
> Let some go a bundle in steam trains that trundle,
> And gey nearly choke you with reek in a tumble;
> A Blue Train for me any day of the week,
> To rocket along like a silent blue streak.
> Airdrie, Balloch and Helensburgh,
> Have waited long and weary for a
> New deal on the railway route –
> It took such a while to come aboot;
> A blue do a'thegither, said some,
> But haud on doubters, here they come!
> It just goes to prove that you don't get a hate,
> That's really worthwhile, unless you wait.
> Where are the queues in the city shops?

Just in the ones that are the tops.
Who wins a fortune on the pools?
Folk who keep filling them up, like fools.
Think on it Mac, – you'd be a dope,
When things are most gloomy to give up hope.
Who knows, some genius may find a way
To fit a transformer to Mr K.,
Replacing dynamic power-blindness
With a wee touch of human kindness.
When money's gone, when folk are thrawn,
Remember, it's blackest before the dawn;
When it seems Fate's dealt a final whack,
Remember, the Blue Trains did come back!
When life goes bang! And all seems lost,
Just soldier on – with your fingers crossed!'

Scoticus – the Saturday Poet

The poem seems to have captured renewed enthusiasm for the Blue Trains and forgiveness for the issues of the previous year. Over the first four-week period, services involving 8,300 trains ran at a level of 92 per cent arrival at their destinations on time. By 31 December they had clocked up over a million miles of service, and minor rectifier faults had been picked up and attended to successfully without significant service interruption. Public reaction to the resumption of services was positive and amongst other publicity drives, BR ran a series of advertisements in the *Evening Times* newspaper that on first glance look like articles; each was headed 'Passengers are people who matter' and contained a brief quote praising the service from a pictured individual. The one for 23 November 1961 featured Fergus Ferguson, a welder in an iron works in Coatbridge, who commuted to work each day from High Street:

> 'I've done it for years by steam train, but what a difference now. The Blue Trains are just the job, clean warm and cosy, especially these cold mornings. Another thing, the timetable suits me fine and the trains are on time and there's plenty of them. I use the Blue Trains at weekends too, down the Clyde to Helensburgh visiting friends.'

Another early convert to the Blue Trains was popular comedian Lex McLean, reported in the *Evening Times* of 12 October 1961 to be commuting from Helensburgh to his show at Glasgow's Pavilion Theatre: 'No traffic worries, no parking problems... just a treat!'

Railway Magazine of December 1961 stated that in the first week, 400,000 passenger journeys were recorded, and that punctuality had broken all previous records. By the end of November this had settled down to a regular 230,000–240,000 weekly passengers paying fares amounting to £17,000, compared with 90,000 and £7,000 in 1960. Again, predictions that the new electric service would bring people back onto trains seemed borne out in practice, but outside railway circles there was little fanfare and scant attention was being paid nationally to this success. The BR house magazine *Transport Age*, whose audience were commercial freight customers, carried a colour feature on the Blue Trains in its April 1962 edition, but this did not enjoy a public readership. There was no British Transport Film Unit short despite their wide coverage of every aspect of British Railways in that era. The in-house BTC academic journal *British Transport Review* contained a paper by the *Glasgow Herald* journalist L.V. Beharrell in April 1962; this discussed the wider impact of the new developments including their detrimental impact on bus services, concluding however that this pointed the way for other cities: 'In Scotland, for the first time for generations, the railway is setting the pace, and showing that hopelessly chaotic and unregulated road traffic need not be the inevitable concomitant of urban civilisation.' [18]

BR's national publicity focus was on the mainline electrification in the north-west of England that would eventually (by 1974) join London Euston Station and Glasgow. Perhaps it was a case of not raising expectations elsewhere – most passenger journeys in 1961 were still being made in steam trains and it would be many years before the Modernisation Plan of 1955 would bear fruit on a widespread basis.

Glasgow Southside Blue Trains

On the basis of the lessons learned with the North Clyde system, work continued to complete the South Clyde electrification (sometimes referred to in information leaflets and timetables as the 'Glasgow Central Line'). A *Glasgow Herald* story on 29 December 1961 announced a satisfactory test run to Neilston from Glasgow Central Station where Driver James Watson of Polmadie Depot was accompanied by Scottish Region General Manager James Ness and several other senior managers. The article also reported that car parking spaces were to be increased at a number of stations to encourage travel by those who might otherwise drive into the City Centre by way of the over-congested Clyde bridges from Southside. The system was ready to open for service by May 1962. Tuesday May 29 saw the first trains run round the Cathcart Circle, to Motherwell via Kirkhill, and to Neilston – a further 27 electrified route miles. The *Evening Times* of 30 May chose to focus its short piece about the reintroduction of the service on a breakdown of one of the new trains that day, causing a very temporary delay, but otherwise indicating events went to plan.

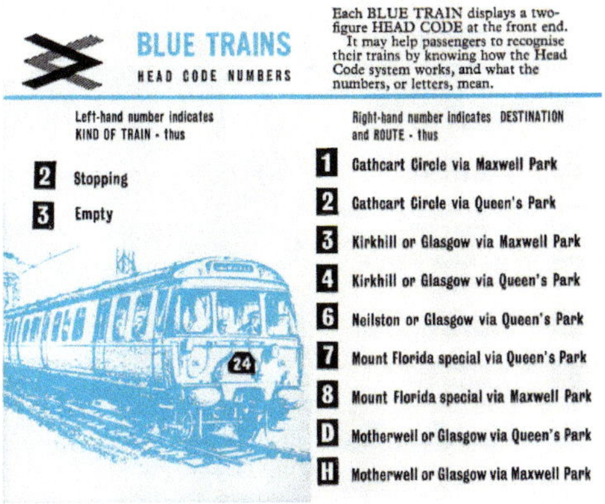

BR South Clyde services head codes leaflet (author collection)

Within the first week South Clyde passenger traffic had increased by 175 per cent, with three times as many people using the service from the middle-class suburbs of Muirend and Whitecraigs. A later assessment of the early years appeared in the *Evening Times* newspaper of 27 July 1965. This reported a 191 per cent increase on the North Clyde lines, with some stations, like Dalmuir Park, showing up to 700 per cent more passengers. On the South Clyde lines, the increase was 200 per cent although profitability was less as these lines had more of a deficit to make up for. The report suggested that both railway staff and users took pride in the Blue Trains to the extent that they were generally tidier and free of the on-board vandalism that plagued electric suburban trains in the London area. Line vandalism, however, was quite another matter as will be discussed below. The picture painted was one of general satisfaction with the services, in contrast with the pervasively negative view of rail travel in Britain. Another sign of popular usage and attraction was the regular appearance of proximity to Blue Trains in house sale advertising in the Glasgow press.

The South Clyde services were different in character to those in the North: a mixture of predominantly middle-class areas within the city boundaries, and Neilston, a dormitory town that had once been a centre for bleaching and dyeing textiles. The line out to Motherwell went into the heart of industrial Lanarkshire with its coal mining villages, steelworks and associated working class communities.

The Cathcart Circle was known for its individual character, celebrated in a 1919 novel *Snooker Tam of the Cathcart Railway*, where it was said

Duchess class pacific on London train with Blue Train on left, Glasgow Central 1962 (painting by Dugald Cameron. Reproduced with permission)

that '... a wag has alleged that this railway was specially built for 'high heid'uns in insurance offices, public hooses, and drapers' shops' No matter! It is a most convenient line and the cheapest thing out of Lipton's.'[19] It was known to thousands who used it to go to the national football stadium at Hampden Park. According to the line's biographer, Jack Kernahan, the Blue Trains robbed the line of its independent character: with the onset of clean trains and, as he admitted, the best service it had ever enjoyed, it seems doubtful the average user in the mid-1960s would have agreed.[20]

One offshoot from the Cathcart Circle route, rare at the time, offered an example of the type of co-ordinated transport recommended by Inglis. In October 1962 a 'Blue Train Bus' was laid on by Glasgow Corporation to take passengers from Castlemilk housing scheme to King's Park Station, offering very cheap through tickets and meeting all trains. A few weeks after the start the *Evening Times* reported that during heavy snow, the Blue Trains on the North Clyde route were barely affected whilst trains and buses everywhere else were delayed and cancelled; these included the Blue Train

Blue Train entering Glasgow Central with steam hauled train from Gourock to right, 1965 (unknown photographer: author collection)

Bus which had battled through from a badly hit Castlemilk to find that trains at King's Park were cancelled. The Blue Train Bus, serving only a few locations in the sprawling housing scheme, was sadly not a success and was withdrawn after only seven months in May 1963 – and is described by one author as having '...soured the Corporation's view of rail-feeders for all time'.[21] It seems that competition between rail and bus services remained a prevalent feature in the early 1960s, fitting with the general policies of Tory government committed to the framework its predecessors had determined 10 years previously with the 1953 Transport Bill – this even though in Glasgow both were publicly owned services. The Blue Trains' success story also hastened the final demise of Glasgow's tram services, the last ones ending in November 1962.

Wemyss Bay and Gourock

On 5 June 1967 electrified services commenced on the Gourock and Wemyss Bay branches from Glasgow Central Station, with total route mileage of

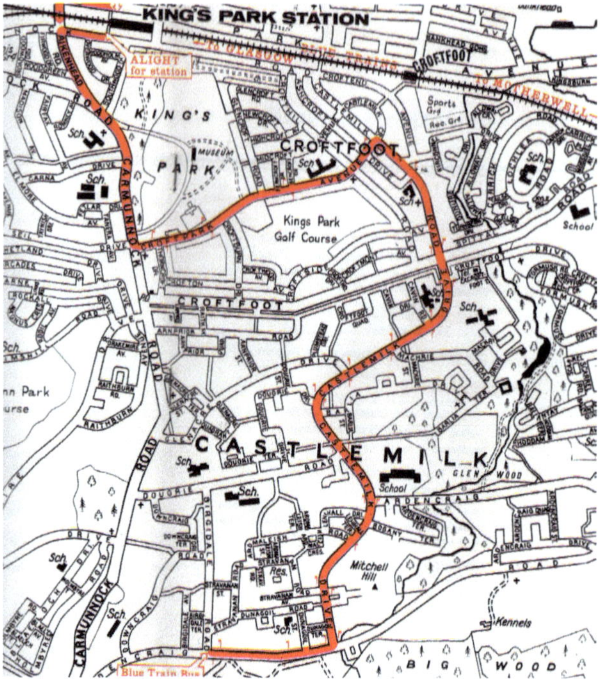

LEFT: Blue Train Bus leaflet 1962 (author collection)

RIGHT: Blue Train bus route in Castlemilk (author collection)

37 miles, extending the total Clyde service routes to 116 miles. The Gourock branch saw the last steam passenger services in Scotland, ending in 1966, so the change was quite dramatic. Nineteen new-build trains, designated class AM11 (later 311), constructed at the Cravens factory in Sheffield, were added to the fleet for the new extensions (Pressed Steel in Linwood no longer built railway stock). These were of similar design to the original Blue Trains but turned out in standard BR blue livery which was now universal. They were fitted with fluorescent lighting instead of the tungsten units in the original carriages and had more powerful motors, which helped on the steeply graded Wemyss Bay line. These lines marked an important departure as they catered for year-round steamer/ferry traffic (at this time these were also run by British Railways) which could expect heavy demand at peak holiday periods. The Helensburgh line included a terminus at Craigendoran for seasonal Clyde steamers but this ceased in 1972 when this pier was closed permanently. Such services had been part of the original visions of Abercrombie and Inglis, but had taken many years to reach fruition.

I am told that the ending of steam on the Wemyss Bay route met with a mixed reaction. Unlike the lines through the city centre tunnels, with their association with grime and discomfort, the steam trains and their carriages – fitted with pictorial scenes over the seats and net luggage racks, evoked

LEFT: BR leaflet for Blue Train introduction to Wemyss Bay and Gourock (Paul Strathdee collection)

RIGHT: Wemyss Bay Station (Dugald Cameron)

memories of a period of travel to Rothesay that was disappearing by the 1960s. Wemyss Bay Station (thankfully!) retained (and still retains to this day) its Victorian architecture that seemed out of keeping with the modernity of the Blue Trains.

Blue Trains in Service... and a few stories

The state-of-the-art Blue Trains were popular with staff as well as passengers. Drivers found them to have good acceleration (top speed was limited to 75mph) and good braking, running smoothly without vibration, and training for the Blue Trains for existing drivers took only six weeks. However, drivers used to steam locomotives were frustrated that when failure occurred there was nothing they could do, whereas a steam locomotive was a creature they held some responsibility for and understood.

Passengers found them very quiet with a noticeable difference when slowing down – the air brakes hissed quietly compared with the noisy steam vacuum brakes of locomotive drawn carriages. They were also generally reliable with maintenance schedules conducted in the new depots at Hyndland and Shields Road south of the Clyde (Hyndland was closed in 1987 and all work transferred to Shields Road).

Over the years, just as in any industry, a mine of stories built up that were exchanged between colleagues and were certainly not shared publicly as they usually involved rule infringement. These reflect the railways as they were then, and are not shared here as any kind of endorsement of the (probably exaggerated) behaviours described. On the Wemyss Bay line, the last train of the day seems to have been the setting of such tales.

One concerns the 'long stretch' between Paisley Gilmour Street and Bishopton – the furthest distance between two stations on the line. One driver (they were all male then) allegedly had a fancy for a particular female guard who must have liked him too. He would invite her into the cab and they would see how far passion might take them before stopping the train again – a journey of less than ten minutes with the male party also controlling the train! The cab on the Blue Trains was expansive and, with the partition in place (or blind down in the days of the see-through cabs), there was room for such dalliances in a manner that might have been difficult in the shared and grubby comforts of a steam engine.

Another story concerns the same late-night train on which back shift railway staff might return home at the end of their working day. One booking clerk boarding the train at Paisley Gilmour Street was spotted by the driver who was a friend, and invited into his cab for the journey home to Wemyss Bay. Cans of 'McEwan's' appeared and were imbibed as the journey proceeded on its way. At one point the driver stopped the train between stations in order to relieve himself, stepping down onto the track to do so. To his companion's horror the train started rolling – the driver quickly finishing his business and catching up with his train to board and resume control – all of which was blissfully unknown to the few late-night passengers. That same booking clerk, allegedly, on another occasion and on his way to work, asked the guard to open the door at Branchton so he could start his duties. The train stopped and the doors opened with the storyteller about to leap out into the dark when the guard shouted ' ....., don't get off, the doors have opened on the wrong side!' The driver, asked by the guard to stop the train at the next station, had inadvertently halted at IBM Halt where the platform was on the opposite side of the train.

Accidents

After the very early problems with the rectifiers and transformers, the Blue Trains were safe to travel and work on. There were accidents but these were not the fault of the trains themselves. The first of these took place on 30 August 1973 at Shields Road Junction in Glasgow when an EMU 303 train from Wemyss Bay travelling at speed collided with the rear of a DMU from Ayr that was pulling away from a signal. The driver and four passengers in the front carriage of the Wemyss Bay train were killed and a fire ensued. On this occasion it was concluded that the driver of the EMU had suffered a heart attack and that the automatic system that should have stopped the train had not had time to activate.

Unfortunately the novelty of electric wires that could sparkle and explode when interfered with, attracted regular vandalism from children in the Glasgow housing schemes through which they passed; this despite effort at education with schools and communities about the obvious dangers at the time the lines were being constructed. The *Evening Times* and *Glasgow Herald* newspapers reported such incidents regularly, including the throwing of objects aimed at the glassed driving compartments of moving trains, and attempts to dislodge the overhead wires and their equipment. One such incident near Easterhouse was to have tragic consequences: on 30 July 1974, a 15-year-old boy, in the company of friends, threw objects at a structure holding up an overhead wire, breaking a strut which was left dangling over the railway line. A short time later a Blue Train collided with this which smashed through the front window fatally injuring the driver, Joseph Conroy, who died in hospital 12 days later. The incident angered drivers who went on strike for a day and from then on, as EMU sets entered St Rollox works for major overhaul, the rounded glass sections were removed and replaced with flat metal-framed front-facing glass sections that offered better protection.

Three other notably tragic accidents involving Blue Trains took place through collisions in what railway people term SPAD incidents (Signals Passed at Danger). The first of these happened on 16 April 1979 in Paisley Gilmour Street Station when a DMU from Ayr to Glasgow was cleared by the guard to leave the station, passing through a red signal as it did so, and colliding head-on with a Wemyss Bay bound Blue Train. Sadly, both drivers and five passengers were killed with many others injured and taken to hospital where three were detained. Lessons were learned and rules changed to try to ensure that this never happened again.

The result of vandalism (*ASLEF 1880–1980* – reproduced with permission)

However, ten years later on 6 March 1989, a similar incident occurred when two Blue Trains collided near Bellgrove on the North Clyde line, one having passed a red light when cleared by its guard (who could not see the signal) to leave the station. On this occasion one of the drivers and a passenger were killed. The rules were reiterated and the layout at the station changed to ensure that signals were visible. Just two years later, a Blue Train collided with a newer Class 314 EMU at Newton Station, killing both drivers and injuring 22 passengers. This was concluded as resulting from changes in the track layout and a missed red signal, and changes were made to avoid future recurrence.

The driver of an empty 303 unit in the tunnel approaching Bridgeton Depot was sadly killed in a collision with an empty 314 unit on 11 September 1986, during shunting manoeuvres. The guard in the rear of the 314 also died from his injuries, and was later held responsible for the accident although instructions and guidance were revised as a result of the official inquiry.

A fatal incident caused by deliberate vandalism occurred late on 25 June 1994 when a 303 unit from Wemyss Bay with six passengers aboard was derailed by concrete blocks placed on the line by two 17-year-olds at the location of what is now Drumfrochar Station in Greenock. The train collided with the bridge collapsing the driver's compartment and killing driver Arthur McKee. Directly behind the driver, passenger Alan Nicol was also killed – having deliberately sat there to avoid the projectiles regularly thrown at passing trains in that area. The two youths received 15-year sentences, upheld on appeal, for culpable homicide. Nowadays railway lines are not so easily accessed so such incidents have reduced, probably also for other social reasons.[22]

As was common practice at the time, the Blue Trains were built using asbestos as fire resistant insulation material. It was cheap, easily manufactured and very effective. What was not widely known until the 1960s was that its dust, caused by any breakage or disturbance, causes cancer and was responsible for many deaths, particularly amongst those involved in its production and industrial use. It has been progressively banned in the UK since 1967 and completely so since 1999. Asbestos is universally present including in homes, other buildings and transport (trains, ships etc.) built before the bans came into effect. Its removal has to be undertaken safely and its disposal strictly governed by law, making this complex and expensive.

Refurbishment and Final Years

With some withdrawals of the original 91 three-car train sets, the remaining 50 of the original class 303 units were refurbished from 1984 onwards. This involved repainting them in Strathclyde orange and black livery windows behind the driver and their replacement with a solid partition. The author has been unable to locate a definitive reason for this last modification, but it is most likely that this was done to avoid distraction for the driver, even though they had blinds that were pulled down at night. The three carriages were now connected by a through corridor, and seating was altered with two plus three seats being replaced by two plus two. The seating upholstery was also replaced with a less attractive design. Train head code panels were replaced with headlights. Importantly, asbestos was removed. The later 311 units were left out of this process, 12 having been redeployed to the London Midland Region some years earlier, and all being withdrawn from regular service by the end of 1990. The diminishing group of 303 sets remained in use for several more years as their more modern replacements came into service, the last 11 withdrawn by the end of 2001 – 40 years after their introduction. In the early 1990s one 303 set (303 048) returned from the London Midland Region, was repainted in the original lined blue and used in service as well as for special occasions. Despite the protests of enthusiasts, it was scrapped after lying idle for a few years at Shields Road Depot. By this time the presence of blue asbestos as fitted to all the original units and still present in 303 048 was rightly considered hazardous and although this looked set to doom all to a similar fate, several carriages were saved; a

Changing liveries – rarely seen in a single train set, Partick 1985 (Paul Strathdee)

TOP LEFT: 1965 BR leaflet (author collection)

RIGHT: Blue Train image still being used for promotion ten years after their introduction (author collection)

LEFT: The Green Train promotion for diesel services, that never really took off – and was anything but green! (author collection)

cobbled-together (with asbestos removed) three-car unit now being at the Heritage Railway at Bo'ness and a motor unit and centre carriage (from 311 units) at Coatbridge undergoing a slow restoration by volunteers.

Last Words on the Blue Trains

Michael Bonavia describes the attraction of electric railways to those concerned with the industry in the 1950s and 1960s, and their likely positive effect on revenue and popularity, as the 'Sparks Effect'.[23] Although this was borne out by the experience in Glasgow, it took many years before this was extended beyond the initial schemes of the 1950s' Modernisation Plan. The success of the Glasgow electrification proved that proper levels of investment and

ABOVE: Ticket for the last Blue Train in service December 2002 (Darrel Hendrie)

TOP RIGHT: The end for one of the Blue Train units (Darrel Hendrie)

RIGHT: The Blue Train logo (BR *Glasgow Electric*: author collection)

attention to detail would bring people back to the railways. This happened also because of the vision, persistence and support of railway workers at all levels: from the senior managers committed to a nationalised railway system as a public service, to the front-line staff who rose to the occasion when the system failed. All this took place under a Tory government in London whose wish to invest in a public railway system was at best ambiguous. The public in Glasgow, many of whom suffered the poorest urban conditions in the UK, took to them with a pride in ownership that was symbolised by the Blue Train label. The major leap in public transport service that was involved cannot be overstated.

The story did not stop in 1967 despite the 1965 British Railways Board Annual Report and Accounts advising parliament that the Wemyss Bay and Gourock electrifications would 'complete the south side of the Clyde Valley Suburban network'.[24] With the most extensive UK suburban rail network outside London, there was much still to do to realise the vision of the wartime planners, and this story continues in the chapters to follow, starting with an account of BR's attempts, often despite rather than because of the ruling government, to make public ownership work.

Chapter 6

FROM BEECHING TO PRIVATISATION

> 'In terms of capital investment for renewal, research development and improvement the railway's share has not been comparable with the millions poured into other forms of transport. New motorways and trunk roads, with all that they mean in environmental offence, between them represent a national investment every year more than five times greater than investment in British Rail, and even then, most of the railway investment is to keep the system going.'
>
> RICHARD MARSH, CHAIRPERSON BRB 1972[1]

Introduction

The Glasgow Blue Trains were the consequence of a combination of planning and policy developments that arose in the post-war period – in a sense the stars aligned, if briefly, to make real a part of the dreams and work of Abercrombie, Bruce, Fitzpayne and Inglis. Although the 1948 nationalisation had a strong socially progressive intention for its socialist architects, the reality, as described in previous chapters, was an emphasis on simply keeping the system going and stemming a haemorrhaging of traffic to make the books balance. The Blue Trains were a beacon of hope for the future amidst general decline. As Richard Marsh (himself a former Labour Minister of Transport) made clear in the statement quoted above, this was primarily the age of the motor car, the only significant rail investment being in the few years following the 1955 Modernisation, from which Glasgow benefited. There were tensions surrounding the need to make British Railways competitive and profitable as soon as the Tory government was returned in 1951, and this gained momentum in the 1960s. This chapter with its focus on national developments examines trends that brought about a much-reduced network but with a modern look.

The chair of the British Transport Commission through the 1950s, Sir Brian Robertson saw the railways as a public service, but his modernisation plans were designed also to make them profitable by 1962. No one at that time discussed railways in terms of their overall social benefits that might override concerns about profitability. This was now to enter the arena with public outcry about the proposals of Robertson's successor, Dr Richard Beeching: these were to drastically cut the network and modernise what remained. The investment that was needed for modernisation was piecemeal and subject to constant change as governments and ministers came and went but despite all this BR was able to produce some notable successes. However, after nearly fifty years of public ownership, it was broken up and privatised, a story covered in this chapter.

The Beeching Era

By 1960 the government were becoming concerned about BR's continued financial losses and a special Select Committee concluded (even though they had difficulty in pinpointing the evidence) that much of the blame lay with branch line passenger services. However, they were quickly reassured by the erudite and persuasive Sir Brian Robertson that he had a plan and all was proceeding well. A small advisory group was set up by the sceptical Transport Minister, Ernest Marples, under the chairmanship of industrialist Sir Ivan Stedeford. This apparently painted a much gloomier picture (its findings and written reports were not placed in the public arena). It concluded that the 'public service' mentality pervading senior management ran counter to the 'business' approach that was necessary to make the railways profitable again.[2] Sir Brian Robertson took early retirement and the structure was reorganised under his successor Dr Richard Beeching, seconded from his senior position in Imperial Chemical Industries (ICI). The changes came into force through the Transport Act 1962, replacing the British Transport Commission (with its wide responsibilities) with the more specifically focused British Railways Board under Dr Beeching's leadership.

Tasked with turning things round financially, Beeching produced his major report in March 1963, causing huge public reaction, especially in the most remote areas of Britain.[3] Most of his attention focused on the closure of most of the country's remaining branch lines, and smaller stations on mainline routes. The NUR reacted with a short pamphlet *The Misshaping of British Railways* in furious reaction to the tens of thousands of their members' jobs that would go (16,200 according to Beeching's initial estimate and later considered to have risen to over 67,000).[4] The NUR asserted that the Beeching Report was bristling with facts and figures that were questionable, and that public opinion was reported by the press to be five to one against implementation. Based on a people-before-profit argument, the NUR stated that the remedy to the losses incurred on the system since Tory government control in 1952 lay in renationalising road transport and ending unfair competition. However, with the exception of a one-day NUR strike in 1962 over workshop closures involving the loss of 18,000 jobs, the unions did not take action against the Beeching cuts. Once it became clear that Labour did not intend to end them, some ASLEF Branches and Districts organised a lobby of parliament in late 1965, but this was not supported by the Executive, who were set instead on making a deputation to the Prime Minister, a long-forgotten and clearly ineffective event.[5]

As well as the thousands of square miles in Scotland that would be bereft of railways, the Beeching Axe fell on suburban lines in the Glasgow area that Inglis and the other post-war planners had considered ripe for electrification. These included the Glasgow Central Low Level route, which like other long-closed lines in the area was subsequently electrified and reopened at great expense many years later. The recent Transport Act had placed the task of putting forward special cases for exemption from closures onto the Transport Users Consultative Committees (TUCCs) and they and public pressure saved some notable routes from the axe on social grounds; these included Scotland's Mallaig, Kyle of Lochalsh and Far North lines. The TUCCs were only permitted to involve members of the workforce (an NUR demand from 1961) in 1964 so until then their perspectives were deliberately limited to passenger hardship. Some

The UK lost a third of its Victorian railway network through the Beeching cuts. This is a typical view of most today (Malc McDonald: Wikimedia Creative Commons)

lines were saved but in certain cases, like the Edinburgh to Carlisle line (Waverley Route) that served areas by then far from any other rail lines, for only a few years. By 1969, after consultations and reprieves, the railway network had been reduced from 18,251 route miles (in 1961) to 12,098; stations had been reduced from 7,025 to 3,002. Staffing was cut over the same period from 500,434 to 275,469.[6] Some of the impact of the Beeching cuts on the network as a whole and its potential for growth would not be seen for many years. By the 1990s the remaining main lines were working at full capacity leaving no room for alternatives that might relieve pressure: the parallel routes considered superfluous by Beeching could have been useful again, as could many of the cross-country routes that were axed, some of which (Cambridge to Oxford, and the partial reopening of the Waverley route in Scotland being examples) are now being brought back to life.[7] In Glasgow, Beeching rationalisation saw the closure in 1966 of St. Enoch and Buchanan St. termini – again an arguably short-sighted decision given the pressure in years to come on the remaining two mainline stations – neither of which could be easily enlarged.

1966 BR leaflet that tried to justify the closure of St Enoch and Buchanan St stations (author collection)

The closures overshadowed the more positive modernisation aspects of Beeching's work, which included the introduction of freightliner services and continued electrification, including the Gourock and Wemyss Bay lines. His tenure also saw British Railways change its name and image to *British Rail*, the busy Design Panel producing new-look logos, uniforms and signage to demonstrate that change for the better was really underway.

Labour at the Helm – Again

The 1964 general election saw Labour ending 13 years of Tory government, no doubt helped along by public displeasure with Beeching's cuts, although there had been no manifesto promise to change or reverse his proposals, and indeed this did not happen. Harold Wilson, the Prime Minister, asked Beeching to devise a fresh transport plan, but, maybe because the end of his secondment was nearing, or perhaps for ideological reasons, he declined and returned to

ICI. His chosen successor was Stanley Raymond, a career transport official (including a spell as Assistant General Manager of BR in Scotland) who had risen to the top from humble origins and had at one time been a trade union activist. He had personal relationships with many senior Labour officials, but had been one of Beeching's chief lieutenants in formulating his proposals.

By 1966, the Government had produced a White Paper outlining their ideas about railways. This accepted that some services (saved from the Beeching Axe after public outcry and rural MP pressure) were now being provided for social reasons and offered proposals to fund them as such. The idea that railways could be provided on the basis that the social costs of closure might outweigh that of keeping them open had been raised in the Beeching Report but his view was that such considerations only arose in the case of suburban services that were economically unviable; in rural areas, he argued, buses could readily replace trains, and in most cases this is what happened.

By the time the Transport Bill went through parliament in 1968, under the Minister Barbara Castle, Raymond had gone following disagreement with her over (amongst other things) industrial relations; she was later to become the architect of the controversial anti-union White Paper 'In Place of Strife'. The changes now separated lines requiring financial support for social reasons from operations that could make a profit; with the writing-off of capital debt, the railways immediately returned to an operating surplus.

The Transport Bill also established Passenger Transport Executives in urban conurbations that would assume responsibility for the co-ordination of passenger services and funding of those regarded as uneconomic but essential. These started quickly in Merseyside, the West Midlands, Tyneside and Greater Manchester; Glasgow, as we shall see in the next chapter, followed a few years later. At last, the idea of co-ordination was established as a policy with the means to achieve its goals, although this would prove problematic in practice. The Labour government had at least inherited some aspects of ideological belief that publicly owned services should be immunised from the ebbs and flows of the marketplace. Bonavia describes the 1968 to 1971 period as a 'honeymoon' under the BRB chairmanship of professional railwayman Sir Henry Johnson, but ending almost as soon as it started.[8] His successor, Richard Marsh, rightly saw that, whilst British Rail was stable, only major investment would achieve the levels of modernisation still required. Marsh was a former Labour transport minister, but is described as having become disillusioned by politics, and having 'turned to managerialism rather than socialism' after dismissal by Harold Wilson.[9] His 1972 Annual Report moved on from the previous model consisting of lists of statistics and figures, interspersed with facts, and into new territory of slick promotion and publicising of achievements. The report talked up the APT project (see below), the nascent channel tunnel proposals, the introduction of computers into management functions and other initiatives including the introduction of ticket machines on the Wemyss Bay and Gourock lines.[10]

Continued Electrification on the Main lines

The prestige project upon which faith was placed for a modern railway was the West Coast Main Line route, with expectation that this would one day go the whole way from London to Glasgow. Work began in 1957 as part of the 1955 Modernisation Plan – the route being considered ideal because of its distance, steep gradients and heavy traffic potential. As noted in an earlier chapter the system chosen in 1956 involved 25 kW overhead electrification. This took place in two distinct phases, with progress never guaranteed due to funding issues and changes of government and transport policy. The first phase concerned the lines from Euston to Liverpool and Manchester, with initial work starting in the Crewe area in 1957 – principally to test equipment and systems. This involved development of all aspects – from new designs of locomotives and EMUs to construction techniques for the overhead wiring. The Glasgow Suburban electrification, by now well underway, both contributed and drew from this process. The lines via Nuneaton to Manchester and Liverpool commenced passenger services in April 1966 and via Birmingham New Street from May 1967. They were immediately successful 100 mph trains cutting journey times – London to Manchester reduced from three and half hours to two and a half. One manager recalled that the 1966 commencement was the only time he could recall the BBC and national press casting British Railways in a positive light![11] It is an interesting footnote to this story that the introduction of diesel locomotives as a stop-gap pending electrification had actually slowed down services – the types available were not as powerful as the steam locomotives they were supposed to replace and did not overcome the labour recruitment issues associated with steam, so many were kept on almost until the very eve of electrification in the North West of England.[12] This issue was tackled for the long Weaver Junction to Glasgow non-electrified section by the introduction of more powerful diesel locomotives (Class 50) between 1969 and eventual electrification in 1974.

Prevarication, however, continued about whether or not to proceed with West Coast route electrification as far as Glasgow. A powerful lobby group, with membership including MPs from all political parties and the National Council on Inland Transport, published a pamphlet in 1969 arguing the case for further electrification, particularly on the Crewe to Glasgow route. This looked in turn at the various alternatives to rail electrification being considered at the time by Government, and at the costs of motorway development, demonstrating that the proposition was cost effective and would contribute to economic progress. The pamphlet was clearly aimed at addressing the vacuum that existed between the very powerful privately owned road lobby, and the virtual absence of one for the publicly owned rail network.[13]

In February 1970, and with no build-up or forewarning, the Government announced that the main line north to Glasgow would be electrified. Work began and was completed in time for an introduction of services in May 1974. Completion also saw short extensions of the South Clyde electrified network to the Hamilton Circle, the branch from the main line to Lanark, the Mossend Freightliner Terminal near Motherwell and some short connecting lines. One of the outcomes was that at last, after 12 years, the North and South Clyde Glasgow electrified suburban lines were joined so that trains could be deployed from one to the other without the necessity of a tow. The overall Glasgow electrified network now extended to 230 miles.

After this, progress in continued electrification faltered. A 1978 BR discussion paper considered phased plans and their costing, suggesting that a further 2,900 route miles (6,800 track miles) could be achieved by 2005, on the basis of five teams completing 250 miles a year in total.[14] This would go much further than has been achieved to date (2026) and would include lines in Scotland to Dundee and Aberdeen. The report compared figures for electrification in other European countries, Britain with 20 per cent of route miles electrified coming near the bottom of nations that had commenced electrification of their railways:

1972 BR publicity pamphlet on the Glasgow electrification (author collection)

		Route Length Electrified %	Total Route Length miles
Switzerland	SBB	99	1,800
Sweden	SJ	62	6,900
Netherlands	NS	61	1,800
Norway	NSB	58	2,600
Luxembourg	CFL	51	200
Italy	FS	48	10,300
Austria	OBB	48	3,400
Japan	JNR	40	13,200
W. Germany	DB	35	17,900
Bulgaria	BDZ	33	2,500
Belgium	SNCB	32	2,500
USSR	SZD	29	84,100
France	SNCF	27	21,600
Spain	RENFE	26	8,300
Jugoslavia	JZ	25	5,800
Poland	PKP	22	14,700
Great Britain	BR	20	11,200
Czechoslovakia	CSD	16	8,100
Hungary	MAV	16	4,700
Portugal	CP	14	1,800
Roumania	CFR	12	6,500

Source: Railway Directory and Year Book 1977.

It should be noted that the NUR were behind the Board with these proposals, issuing their own pamphlet on the matter. They lobbied Government for the necessary investment through its parliamentary group of MPs: this, by now, after systematic effort, had grown to 11 Labour members, and increased to 12 in the 1979 general election. Harking back to the Weir Report of 1931 (see Chapter 1), a December 1978 NUR submission to Government looked at the then recent oil crisis, its rising cost, its environmental impact and the likelihood of supplies eventually running out, to talk up the benefits of large-scale investment in electrification. It offered vigorous support for 'Option C' – the ultimate one considered in the BR discussion paper referred to above, but believing that this should be completed by 1995 rather than ten years later.[15]

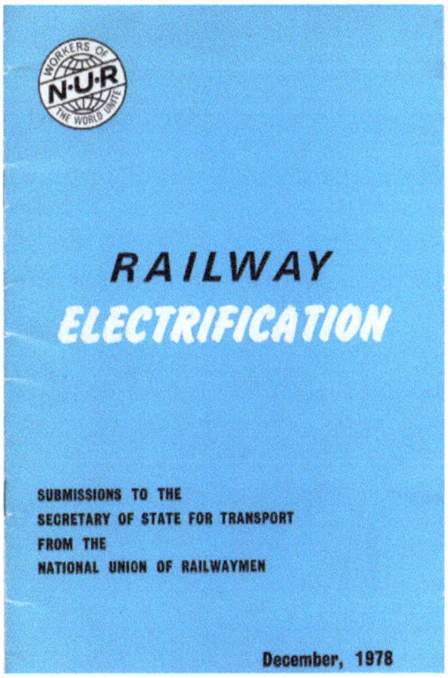

1978 NUR pamphlet calling for further electrification (author collection)

ASLEF also argued proactively for such investment: in a 1982 lecture, General Secretary Ray Buckton suggested that even the medium option for widespread electrification offered in the 1978 paper would have cost the same as the new motorway between Birmingham and Oxford. This even though a case for the new road had not been made, and announced just a week after BR were refused £10 million to update commuter services in London and the South East.[16]

InterCity, APTs, Jimmy Saville, HSTs – and the end of BR

By the 1970s consideration in other countries was being given to the building of new straight lines that could cope with high-speed electric and diesel passenger trains – High Speed Trains (HSTs). In Japan, the first Bullet Train services were introduced in 1964 – specially built straight lines that joined urban areas. Historically, the late and faltered development of railways in Japan that did not suit high speeds enabled this investment and its continued development right up to the present day. In Europe, Italy pioneered the first purpose built high-speed line between Rome and Florence which opened in 1977. Since then, most Western European countries have followed, with an increasing high-speed network linking countries as well as the cities within them. In some cases, this involved new built lines, in others, straightening out curves. In Britain, other cheaper solutions were developed based on the premise that Britain's Victorian-built railways commonly included tight curves (50 per cent of the mileage) which would be prohibitively expensive to iron out. The country's small size and population density ruled out the building of dedicated new lines (a theme controversially revisited in more recent times with the much-curtailed HS2 development discussed in the last chapter).[17]

The high-speed process started with the concept of the Inter-City train, a modern version of the express trains that had been around since the 19th century – many named and famous like 'The Flying Scotsman' between London and Edinburgh. First notioned in the Beeching Report, the Inter-City idea (which became the 'InterCity' brand) was developed in the 1960s to popularise high-speed diesel and electric-powered trains linking Britain's largest cities. By the late 1960s

Top: The new look BR InterCity logo from the 1972 Annual Report (author collection)
Bottom: Travelling in comfort in 1972 – from the BR Annual Report (author collection)

and early 1970s continuously welded rail and powerful diesel and electric locomotives were taking people between London and Edinburgh by diesel haulage at speeds of up to 100mph, and the same on the electrified West Coast Main Line to the cities of the north-west, and eventually Glasgow. New non-corridor coaches with higher standards of comfort were introduced for these services.

The novel approach to Britain's Victorian age railway problem was to seek an engineering solution through the development of tilting trains that would float round curves in rather the same way that a racing motorcycle leans to take corners at speed. This, it was envisaged, could enable travel at speeds of up to 155 mph. However, the concept would take time to develop so in the meantime a new class of diesel passenger unit was introduced that could travel at speeds of up to 125 mph – the InterCity 125s. Ninety-five sets, built by BR Engineering, were brought into service between 1976 and 1982, operating across Britain and considered highly successful. Whilst journey times were cut significantly, their popularity was at least partly the result of heavy advertising, fronted by the then national treasure Jimmy Saville: who around in that era can forget the slogans: 'This is the age of the train' and 'Let the train take the strain'? After his death, Saville was exposed as a serial sex abuser to the extent that would have brought him a long prison sentence, but happily the InterCity 125s outlived their association with him and their popularity continued for many years. Their success in speeding up travel was helped by engineering improvement – the curvature on some routes was reduced.

No InterCity 125s were allocated to Scottish internal services until 2018: 26 redundant but refurbished sets from the west of England after, in some cases, 40 years of service. Their replacement by Scottish Government owned ScotRail was announced in 2024, but they remain in service on the non-electrified lines from Edinburgh and Glasgow to Aberdeen and Inverness at the time of writing. It would be fair to state that although now well past their sell-by date and lacking some of the safety and environmentally protective factors now built into modern

1976 BR publicity pamphlet for its new diesel-powered high speed trains (author collection)

trains, the InterCity 125s enjoy similar icon status to the Blue Trains – amongst both railway staff and the public who used them. Things had come full circle, if in reverse – having been the first to enjoy the modernity of the Blue Trains, Scots were now last to get the benefits of the InterCity 125s – after everyone else had finished with them – of course under a very different, fragmented and a mostly privatised network regime.

Meanwhile work was underway in the BR Derby research and design centre on developing the tilting Advanced Passenger Train (APT) that would obviate the need for newbuild railway lines. Starting in the late 1960s, the first gas-turbine powered prototype unit appeared in 1971 and began making test runs on a designated 13-mile section of redundant track in Leicestershire in 1972. However, it was immediately subject to boycott action (and a one-day national strike) by ASLEF as it only required a single driver and had no capacity for a second operative (soon altered). The pause was used to iron out technical issues and further testing on several mainline routes broke speed records over the next few years, including a record 152.3 mph between Swindon and Reading in 1975. In 1979 the first complete electric APT for the West Coast Main Line appeared, the first of three units, and after further trials, was launched with a press run in 1981. The press reaction was poor, with reports of nausea and sickness as the train went round bends on its journey southwards from Glasgow. Without any fanfare two of the further modified units successfully entered service in 1984 between London and Glasgow but were withdrawn in 1986 and the project abandoned. By this time the InterCity 125 trains had proved their worth on many routes and were seen as the future rather than the tilting APT.

BR publicity postcard for the tilting APT service from London to Glasgow late 1970s (author collection)

The plans for the tilting train developed by BR were used by Fiat-Trentino in Italy (using bought APT patents and technology they had been developing themselves) leading to the very successful introduction of Pendolino tilting trains throughout Europe over the next few decades. In the early 2000s these were reintroduced onto the West Coast Main Line by private operator Virgin Trains – the adverse reaction from the 1980s long forgotten, and the technology expensively reimported to its birthplace. These were initially limited to 125 mph but continued track improvement led to faster speeds and a special speed record attempt in 2021 between London and Glasgow – which failed to beat the one set by an APT in 1984 by 20 seconds.

The concept of the specially built High Speed Train (HST) line did not go away. The Channel Tunnel that originally opened in 1994 involved the building of the UK's first HST link in 2007 – from St Pancras Station in London, 69 miles to Folkestone in Kent, with electric trains from London to Paris and Brussels running at speeds of up to 186 mph. With this line known officially as 'HST 1', attention turned to 'HST 2' – a projected new link from London to Birmingham and thence northwards. This as we shall see later in this book was subject to the same investment issues and political uncertainties as those experienced by Richard Marsh in the early 1970s.

British Rail ceased to be in 1994, but in its final years achieved some notable achievements: InterCity 125 HSTs – the fastest diesel trains in the world; successful upgrading of the existing railway system to achieve competitive high-speed travel; operation at lower levels of public subsidy than most European countries; a fourfold increase in productivity during its existence in terms of passenger miles per employee (some of which was achieved through socially retrogressive measures by Beeching); the popular InterCity brand emulated in other countries but ditched on privatisation in 1994; the introduction of market-pricing fare structures that made fares affordable; 11,000 route miles that were in substantially better condition than when it took over in 1948; tilting train technology – proven over time if not developed by BR; modernisation of freight services e.g. freightliner and merry-go-round systems; airport links across the country; research and development expertise that was sold elsewhere; and the introduction of electronic signalling.[18] Not forgetting of course, the highly successful Blue Trains! Such achievements went largely unrecognised at home due to a persistently poor public image typified by the concocted comic myth of the 'British Rail Sandwich'.

Tory Governments and Privatisation

History is now passing judgment on the rush to impose an economic model on the country that was founded on the belief that market forces rather than government ownership and intervention would self-regulate and lead to general prosperity. Privatisation is not looked back on favourably, being associated with the decline of manufacturing and an associated rise in service and financial industries centred on London, mass unemployment, increasing inequality and social dislocation. In some cases, like water supply and sewage disposal in England and Wales, the outcome has been completely disastrous as profits have gone to shareholders out of all proportion to the requirement to maintain and update infrastructure. Whilst the rush in this direction took place during Margaret Thatcher's period of government in the 1980s, she resisted privatising the railways wholesale, instead selling off peripheral parts of their business to sink or swim under private ownership: hotels, shipping, catering and most significantly British Rail Engineering Limited (BREL), the engineering subsidiary that provided workshop and rolling stock production services and was already subject to a compulsory tendering process. This was a core railway activity and the fact that the sale did not attract much interest points to its ultimate decline and contraction as an industry so that today no locomotives are built in the UK, only the construction of passenger-carrying multiple units dependent on their overseas owners for component supplies.

The idea of full privatisation carried over into the 1990s and the successor Tory administrations. This process was pushed along to its eventual conclusion by European Directive 91/440 that ruled that rail networks should separate track infrastructure from the trains that used them so that charging facilities would enable open access. This model might arguably have suited mainland Europe where cross-border services were fundamental to travel, and companies remained in public ownership. In Britain where this did not apply (at least until the Channel Tunnel was opened), the prevailing political climate could only lead this idea to network fragmentation and privatisation. The irony here, as most commentators have noted, is that BR was beginning to work well and silence its critics by improved and faster services, but ideology won the day with little informed debate about the issues involved.[19] There was discussion within the cabinet about the form their 1993 manifesto commitment to rail privatisation should take, with Prime Minister John Major's view prevailing: that in order to maximise revenue, multiple companies would be created who would be awarded time-limited franchises whose funding from Government would depend on contract compliance and performance. The tracks and infrastructure would be provided by a single national organisation *Railtrack*, who would lease stations (with the exception of a small number of mainline termini including Glasgow Central Station) to the franchised companies, and charge for use of the tracks. By the end of the 1990s what had been a single entity prior to Thatcher's ascent to power, British Rail, had been broken up into approximately 100 companies, with competition and profit, rather than public service, as the main driver. Whilst the main lines and regional lines were broken up into Train Operating Company (TOC) franchises that have changed hands several times since the mid-1990s, in Scotland the trains were run by a single company, ScotRail, whose story under privatisation will be told in the last two chapters.

Privatisation was a failure, demonstrating all that was wrong with marketising and fragmenting a vital public service. Railtrack was the only piece of the system to be put on the stockmarket (in 1994) as it was clearly seen as

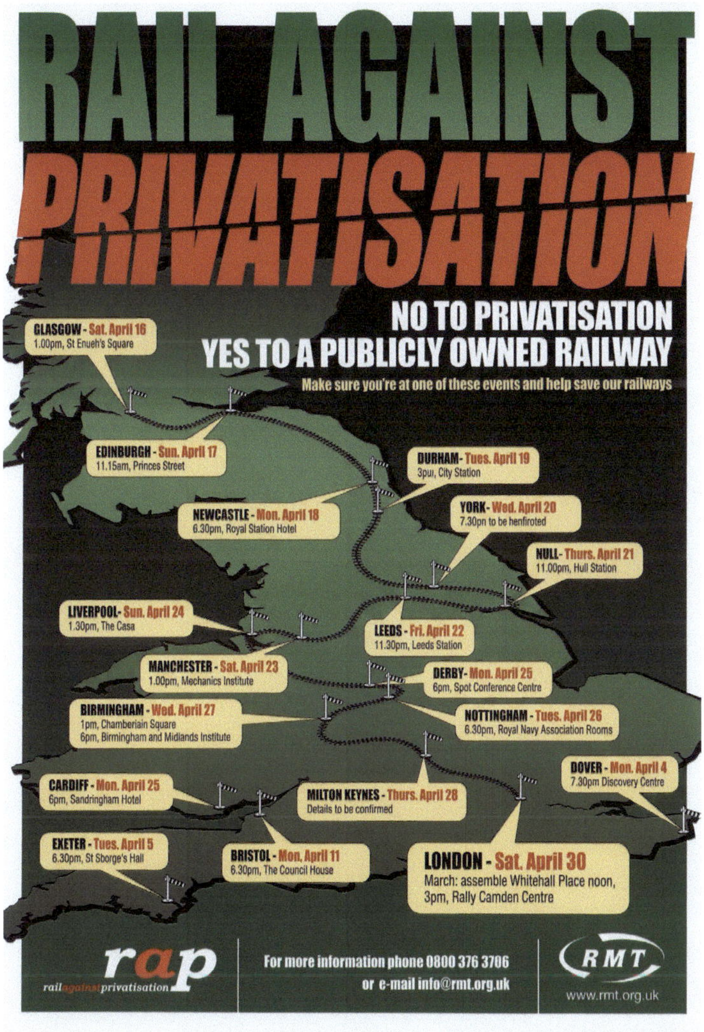

Campaigning against rail privatisation mid 1990s (author collection)

the gravy train for investors for whom it would suck up revenue from the TOCs – its board and managers reflected that priority and became a watchword for arrogance and maladministration. When profit becomes the prime operating motive, investment inevitably suffers and this was certainly the perception surrounding a series of major accidents: Southall in 1997, Ladbroke Grove in 1999, Hatfield in 2000 and Potters Bar in 2002. In each case safety had been compromised because of the break-up of the national railway system. Privatisation and the dispersal of the workforce had, amongst other things, casualised much of the track maintenance workforce, breaking the links and relationships between staff on the ground: signal staff in rural areas where speeds might be high and risks increased would not know the staff working on track maintenance or trust their knowledge of layouts. Automatic Train Protection (ATP), an advance on long-established safety systems designed to prevent accidents through Signal Passed At Danger (SPAD) incidents, eliminated driver error or negligence (accidental or otherwise) by automatically applying braking. This was in the process of trial at the time of privatisation, governed by the previous cost-benefit analysis system (known as Value per Prevented Fatality or VPF) that deemed it a worthwhile safety investment as it might save lives. The new companies, abandoning VPF, preferred a cheaper system and were only prepared to pay for ATP if the Government stepped in with funding.[20]

Railtrack was re-nationalised in 2002 and renamed *Network Rail*; it was failing financially, and urgent government investment was necessary to address the vast number of speed restrictions that had been put in place as a result of major accidents. Railtrack's performance had effectively wrecked the system across the country and, with fragmentation now universal, undone much of the progress made in the last years of British Rail; this despite lack of Government investment when compared with systems elsewhere in Europe. Since then, several of the TOCs have been taken back into public hands due to their failings. The Labour government elected in 2024 resolved to bring the remainder back into public ownership as franchises came up for renewal, with a rebranding of the system as *Great British Railways*. This latest episode in railway history is not a return to the British Railways of 1948: the TOCs will be gradually incorporated into the new organisation, but rolling stock will remain in private hands and be subject to continued tendering, contract and lease arrangements; capacity to build locomotives and train sets seems unlikely to return to public ownership any time soon.

The next chapter will take us back to Glasgow and developments with the suburban transport system from the 1970s onwards.

Chapter 7

LATE 20TH CENTURY PLANS FOR GLASGOW

> *'Taking account of the increasing demand for travel and recognising that unrestrained road traffic demand cannot be catered for simply by building new roads, plans are afoot to provide the modern equivalent of trams, a light rail system, utilising the versatility of modern technology to reach areas presently not well served by rail.'*
>
> STEVEN LOCKLEY, DIRECTOR GENERAL, STRATHCLYDE PASSENGER TRANSPORT EXECUTIVE, 1990[1]

Introduction

The last chapter offered the broader story of electrification and modernisation of railways at national level in Britain after the 1960s, based upon the policy context. My account centres on a belief that the piecemeal nature of electrification demonstrates that progress was always a hostage to fortune determined by wider considerations, and not (as it could and should have been) continuous development of essential railway service and infrastructure. Notwithstanding this background, and sometimes against the odds, BR managed to achieve success in a number of areas, but this was halted and set back by a privatisation process in the 1990s that never settled down.

One policy aspect that did have an impact, at least for a while, was the enablement of regional transport executives through Labour government legislation of 1968. This chapter focuses on how this resulted in developments in the Glasgow Region, involving highs and lows, missed opportunities but hope for change through the efforts of planners, activists and policymakers. The chapter takes the story up to the turn of the century and devolution of government in Scotland.

The Greater Glasgow Transportation Study

In 1964 Glasgow Corporation, BR, the Scottish Development Department and the Scottish Bus Group came together to initiate a study of transport usage in the Glasgow area and forward plans to 1990 – *The Greater Glasgow Transportation Study* (GGTS).[2] With an eye to the USA and solutions to urban transport issues that were on the policy agenda, a firm of New York consultant engineers and architects were engaged to lead the study – Tippets, Abbett, McCarthy and Stratton. They arrived in Glasgow in 1964 complete with a young team of American researchers, and began an extensive survey based on questionnaires to the public. When the final report was published it was said to be the first such exercise of its kind in Britain. However, what it lacked was any gathering of public views, either by questionnaire or consultation exercises, so the question of public support or otherwise barely came into the equation, despite Glasgow Lord Provost John Johnston's promise that the outcome would give citizens '... the kind of transport system which they need, want and deserve'.[3] The report produced a vast amount of data that was used to forecast needs using other information sources including population growth and car ownership predictions – both underestimated, as it eventually turned out. It also had to work around plans already in place, notably the Wemyss Bay and Gourock railway electrifications, and road-building programmes. Whilst the railway electrification was based on proven experience and popular demand, the road construction involving 70 miles of motorway and expressway can hardly be said to be so – especially considering Glasgow's relatively low levels of car ownership. Whether the population needed it or not, this included the M8 through the centre of Glasgow over the Kingston Bridge (completed 1970) and the Erskine Bridge over the Clyde downriver from Clydebank (completed 1971). This road building had its origins in the Abercrombie Report of 1945, which had also given birth to the Clyde Tunnel between Whiteinch and Govan (opened 1963). The report recommended further phased motorway and expressway development – north towards the Trossachs, out to Ayrshire, to Johnston and between Paisley and Hamilton, most of which was realised by the end of the 1980s.

Rail Service

Electric, diesel and some steam powered trains are operated in the Greater Glasgow Area by British Rail. Numbers of electric and diesel power units and unpowered passenger coaches in service as well as seating capacities are shown in Table V-16.

Table V-16. Railway Equipment

Equipment	No. of Units Electric	Diesel
Power Units	91*	105*
Unpowered Coaches	182	195
	(80)**	(55)**

*Includes power units with passenger accommodation

**Average car seating capacity for multiple units

Figure V-23 shows the rail routings in the Study Area by type of service and equipment used (electric or non-electric). Electric equipment operating out of Queen Street Low Level and Central Stations is the most modern in service in the Greater Glasgow Area, and railway services from Glasgow to Gourock and Wemyss Bay are presently being electrified.

Railway stations in the central area are Central Station, St. Enoch Station, Queen Street Station (Low Level), Queen Street Station (High Level) and Buchanan Street Station. Figure V-24 locates these and other rail, underground and bus stations in the central area of Glasgow. (Since the survey was conducted in 1964, St. Enoch Station has been closed and facilities at Queen Street Station (High Level) have been modernised.) Numbers of trains entering the central area daily and in the morning peak hours are shown in Table V-17.

Table V-17. Rail Service at Principal Stations in the Glasgow Central Area

	Suburban Trains		Long Distance Trains	
Station	No. per day	No. in a.m. peak hour	No. per day	No. in a.m. peak hour
Central	273	25	41	5
St. Enoch	53	8	68	11
Queen St. (Low Level)	324	25	-	-
Queen St. (High Level)	-	-	43	3
Buchanan St.	-	-	26	3
Total	650	58	178	22

Data analysis from the 1960s *Greater Glasgow Transportation Study* illustrating the rapid development of electric services (SPT Library)

The analysis of existing information in the report showed starkly that whilst railway lines previously recommended for electrification had closed, those that had been electrified had seen dramatic increases in passenger use. The report's intention was to turn this around and plan the phased development of an integrated transport network in the region. Whilst based on a premise of increased car ownership, it also believed that transport systems should be improved to attract usage before people turned to car ownership, rather than trying to woo them back when it was too late – a feature that would plague developments in the years to come.

The GGTS concluded with grand recommendations for railway improvement, much of which went back to Inglis and earlier: the Glasgow Low Level line closed only a few years previously would be reopened and electrified between Stobcross and Rutherglen; the line to East Kilbride would be electrified with a new station under the town centre; a new station at Motherwell; the closed suburban lines through Maryhill in the north of the city would be reopened and electrified; the line from the north of the city to

Cumbernauld would be electrified with a reopened and electrified branch to Kirkintilloch; the North Clyde electrified line would be extended a short distance from Airdrie to Drumgelloch. The plan involved the reopening of 18 stations and the building of 14 new ones with 196 additional miles of diesel-powered and electric railway. Total costs for full implementation by 1990 (in four phases) would be £62.7 million (approximately £163 million today) in rail investment (including the Wemyss Bay and Gourock electrification already underway), and, tellingly, £388.2 million (over one billion pounds today) on road development, much of which had already been committed when the study commenced.

What was absent from the GGTS proposals was any mention of tram/light railway/rapid transit systems – a marked departure from the vision of Fitzpayne (who retired in 1969) and the other post-war visionaries. The Glasgow Subway did receive early funding agreed through GGPTE in 1974 for upgrading. It closed for two years in 1977 for extensive work to be undertaken, involving rebuilding of stations, repairs to the twin tunnels, replacement of all the antiquated equipment and new train sets – at a cost of £43 million (approximately £293 million today). The Light Railway Transport League, who had published proposals on several occasions during the twilight years of the Glasgow trams, reacted to the 1974 announcement with a plan of their own – *New Life for the Subway*. This involved extending the 4 feet gauge system around the north of the city, leaving the existing circle at Kelvinbridge, and rejoining at Renfield Street in the city centre. The imaginative aspect of their 'Northern Circle' proposal was that it took up the abandoned 1967 GGTS suggestion that would reopen closed BR track beds. GGTS had proposed standard electrification of the lines, but the Light Railway League plan was for a cheaper light rapid transit railway – altogether 1.25 miles of existing subway (between Renfield Street and Kelvinbridge), 6.25 miles of closed BR lines, and 1.25 miles of new construction – 8.25 miles in total serving 14 stations from Botanic Gardens through Maryhill, Possil and Stobhill and thence turning south through Red Road (a densely populated council development of multi-storey flats) to Renfield Street. The League, whilst not costing their scheme, did a survey of public opinion drawing a positive response, and also surmised that such a development could increase the popularity of the subway and reduce traffic congestion in the city centre through its park and ride facilitated stations.[4] Whatever its merits, and they seem obvious, and drawbacks (such as its use of an obsolete gauge that would always restrict access and use), their proposal, as with so many before and since, fell by the wayside.

Greater Glasgow Passenger Transport Executive and Trans-Clyde Developments

The publication of the GGTS report in 1968 coincided fortuitously with the Transport Bill 1968 which facilitated the establishment of passenger transport executives in Britain's metropolitan areas. Such a body for the Glasgow region could co-ordinate implementation in a manner found problematic in the previous attempts immediately after the war. Although announced for the Greater Glasgow area by Secretary of State for Scotland Willie Ross in February 1970, this looked like a stillbirth after the Tories were elected to Westminster in 1970. They were expected to stall plans for PTEs until previously agreed local government reorganisation had taken place, a position adopted by the Tory administration in Glasgow Corporation amidst warnings by MP Teddy Taylor that a PTE would drain the city's funds through support for rail services. However, the go-ahead was given by the Tory Secretary of State Gordon Campbell in 1971 for an interim authority pending local government reorganisation when this would become a regional responsibility within the new two-tier system. This would give the large regional authorities (in this case Strathclyde) responsibility for transport co-ordination, and take this away from Glasgow, whose new city council would have reduced responsibilities.

The Greater Glasgow Passenger Transport Executive (GGPTE) began operating in October 1972 taking in a vast 662 square mile area encompassing the urban populations surrounding the city itself. Under Section 20 of the Transport Act 1968, GGPTE now became responsible for funding all the services run by BR in its area, although the burden was eased by financial support from the Scottish Office for the initial period, reducing annually from 85 per cent in 1975 to zero in 1979. Change was immediate: the 'GG' logo appeared on trains and city buses, and some feeder 'Interlink' bus services were introduced to rail stations throughout the area. Most significantly, the GGTS proposal to reopen the line through Glasgow Central Low Level was taken up immediately as the 'ClydeRail' plan: a case had been put to Government for funding for the project by British Rail in 1971, and this was revamped as a GGPTE proposal in May 1974 – and agreed by the Secretary of State.[5] The new partnership carried forward these developments and added ideas of their own, the most significant being a new burrowing junction at Kelvinhaugh which allowed eastbound trains to access the Glasgow Central Low Level line without crossing the path of those emerging from Queen Street Low Level. A link from the south was also proposed at St Johns, which would

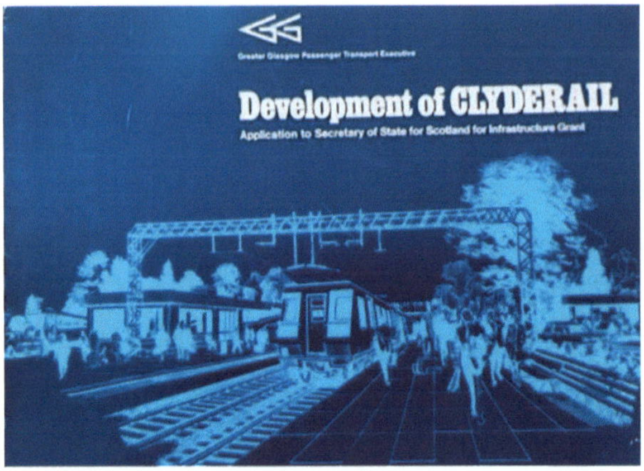

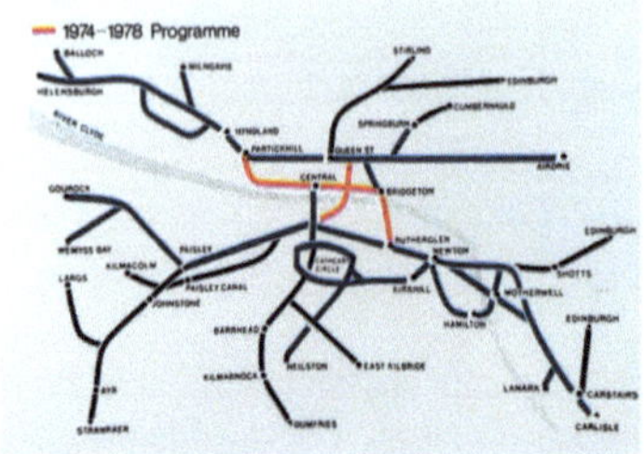

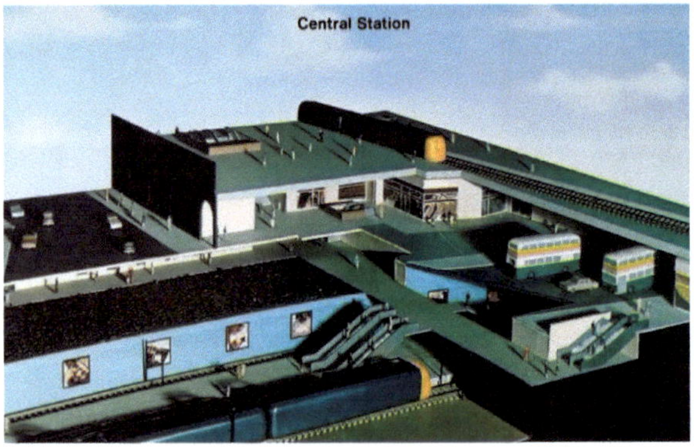

TOP LEFT: Strathclyde's 1974 funding bid for Clyderail development (SPT library)

ABOVE: The proposed improvements in the 1974 report (SPT library)

LEFT: A graphic showing the proposed reinstatement of the Glasgow Central Low Level line (SPT library)

have enabled electric trains from the Paisley direction to run through to Airdrie, but this and some new station developments were ultimately dropped on cost grounds. The cost in 1974 was estimated at £26.57 million (over £255 million today).

The report also took account of the need to target developments towards the region's inequalities that were characterised by its many areas of multiple deprivation, a commitment that would shape overarching policies of the Strathclyde Region from 1975 onwards. By now the mistakes of the post-war planners were becoming glaringly apparent, the peripheral housing schemes and multi-storey flat developments harbouring social problems that had become a priority for attention.

Work on ClydeRail, subsequently re-branded as *The Argyle Line*, commenced in July 1974. Construction involved extensive engineering to allow clearance for the electric wires through the 4.77 miles of tunnel which were laid with concrete slab track with the surface paved in 300-foot sections. The burrowing tunnel involved 250 metres of new tunnel and 450 metres of

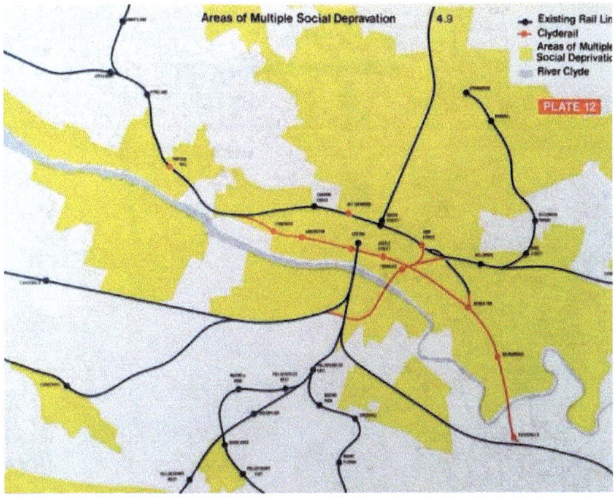

Left: Map from the 1974 report showing areas of deprivation and their rail links (SPT library)
Right: The Red Road Flats in 1978 (© Virginia Turbett)

cutting; between August 1976 and January 1978 this aspect included demolition and careful monitoring of the impact of the operation on surrounding buildings including a primary school.[6]

Opportunity was taken to use the contractor work teams to replace the remaining 6.25 kW sections on the North and South Clyde lines. A new station was built in Partick (an interlink with the Subway and buses), and stations were reopened at Rutherglen, Bridgeton, Anderston, Stobcross (later renamed Exhibition Centre) and Central Low Level. A completely new station was built underground in Argyle Street in the city centre, replacing the original one nearby at Glasgow Cross. Sixteen new three-car EMUs (Class 314) were built by BREL in York for the service, with air-operated doors and passageways throughout. These, I am advised by former railway staff, were cheaply built and simply not as good as the Blue Train units. They were, however, long lasting, with their final withdrawal taking place in 2019 as they did not comply with access requirements for disabled persons.

The new Argyle Line was completed towards the end of 1978 and opened by the Queen with great fanfare in November 1979, alongside the refurbished subway (open to the public a few months later). All trains were now liveried in corporate orange, the blue and white that had replaced the original Blue Train livery being phased out. The whole suburban system, including the revamped subway, was now rebranded as *Trans-Clyde* and there was considerable effort to sell this as something innovative, new and attractive. City centre dwellers were told 'If you live anywhere in or around the centre of Glasgow, welcome to the ultra-modern Trans-Clyde public transport

Integrated transport – from 1979 SPT Trans-Clyde promotional material (author collection)

system!'. The same brochure quoted BRB Chair Sir Peter Parker when he switched on the 25 kW system at a ceremony in the summer of 1979: 'This scheme is part of a grand design which will bring easier travel to the citizens of Strathclyde. And this symbolic gesture is a shining tribute to the men who transformed a closed railway line and derelict stations and tunnels into facilities fit for today and the future.'[7]

The first Trans-Clyde timetable 1979 (author collection)

The introduction of the Argyle Line resulted in changes to driver recruitment and training. The traditional staged and graded training route for staff who had been termed 'enginemen' had its origins in the steam age, and it was replaced with a shortened system that turned out drivers from scratch over a 12-month period – these new staff known derisively by their predecessors as 'Bridgeton babes' and 'boil-in-the-bag' drivers.

Strathclyde Passenger Transport Executive and the Recession of the 1980s

Whilst work was under way on the Argyle Line, GGPTE became Strathclyde Passenger Transport Executive (SPTE) when Strathclyde Regional Council took over responsibility for transport upon local government reorganisation in May 1975. Although Strathclyde covered a vast area – 4,726 miles with a population of 2.5 million, the SPTE boundaries did not change, based as they were on the parameters set by the 1968 legislation. This resulted in the peculiar situation of the Region having a wide responsibility for half of Scotland's population including meeting their transport needs, but tied to making Section 20 subsidies within the PTE's boundaries around Glasgow. Beyond them railway finance was the responsibility of BR. This was bound to complicate future rail development as most lines and services ran beyond these boundaries, some to well-populated areas, particularly Inverclyde (Renfrewshire) and Ayrshire.

The 1979 opening of the Argyle Line and the realisation of the concept of Trans-Clyde marked the high water of SPTE's efforts to integrate transport in the region. From that time on it was all downhill as they were forced into defensive reaction to the central Government embrace of the market and relaxation of control and regulation.

In December 1983 SPTE, Strathclyde Regional Council and BR published the findings of a significant review into the region's rail network.[8] This was very much influenced by the economic recession of that period, the expansion signified by the Argyle Line opening being followed quickly by warnings of closures and contraction, especially on the many diesel-operated lines in Strathclyde which were often typified by poor levels of service and heavy losses. In January 1983 the Paisley Canal to Kilmacolm line had finally closed after years of uncertainty and cuts in service. Buses provided by private operators were instead offered to travellers from the Glasgow suburbs of Corkerhill, Mosspark and Crookston as well as out beyond Paisley to Kilmacolm. Passenger use was falling on all the electric services too – from 48.6 million journeys in 1980/81 to 43.6 million in 1981/82. This reflected closures of workplaces and reduced household spending generally – factors probably more significant in the region than growing car ownership. Glasgow and Clydeside generally were de-industrialising at a very rapid rate as government policies took effect. Electric services on many routes were cut as the strain was felt on the Section 20 budget. This resulted in six of the 303 EMUs being transferred to the London Midland Region's Crewe area. The 1983 review tried, however, to offer some optimism: the go-ahead had already been given in 1982 for the electrification of the busy lines (totalling 52 miles) from previously electrified Paisley to the Ayrshire coast at Ayr, Ardrossan Harbour and Largs. This was enabled by grants from Government (no doubt helped by the lobbying effect of Ayrshire's Tory MPs), Strathclyde Region and the European Union, with a total cost of £81 million. The lines were all electrified by 1987 and promoted through a rebranding as *AyrLine*.

While the review admitted the absolute importance of rail assets for economic development, it looked at how a reduction in BR's Section 20 claim from £29.8 million in 1983 to £18.6 million in 1988 might be achieved. This would involve '... a balanced package of cost reductions, improved productivity and investment in modern equipment.' The loss of traffic and revenue was exacerbated in Strathclyde by the expensive requirement to remove asbestos from the fleet of 303 EMUs. The good news in the review included plans for electrification of lines to Cumbernauld and East Kilbride, some reopening of a few long-closed stations, the refurbishment of others, the expansion of services from Glasgow to Kilmarnock and the reinstatement of services from there to Ayr. The bad news was that the Neilston line would be closed beyond Whitecraigs. Refurbished stations and modernised operational practices would involve staff redundancies with the closure of most booking offices and the introduction of ticket machines. Most controversially, suburban and regional trains operated by EMUs and DMUs would become Driver Only Operated (DOO) progressively between 1985 and 1990. In all some 500 rail jobs would go.

The DOO proposal was a red flag to the rail unions and disputes surrounding it de-railed and eventually altered the 1983 review's proposed timetable and implementation. DOO meant that drivers would remotely operate the electric doors that existed on the newer train, making guards progressively redundant on more and more services as the older DMUs were replaced. This was not a situation unique to Strathclyde as similar

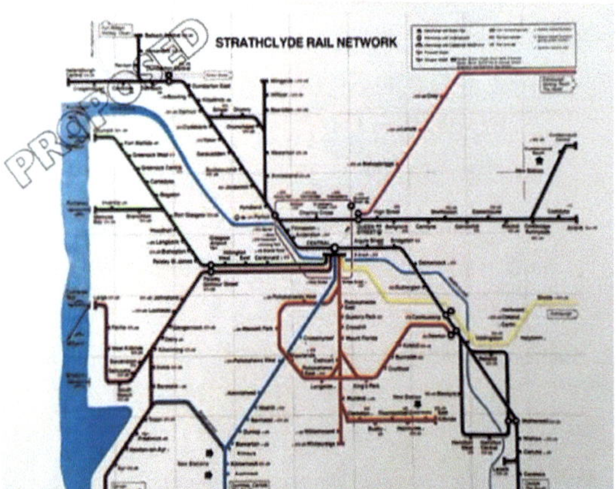

LEFT: Hesitant progress: the 1983 Rail Review (SPT library)
RIGHT: Modest proposals in the 1983 report (SPT library)

confrontation was taking place elsewhere in England and Wales and was the subject of prolonged negotiations at national level between BR and the unions. This was a period of heightened industrial relations tensions generally against a background of the momentous 1984–85 year-long miners' strike. The National Union of Mineworkers (NUM) faced a Tory government determined to beat them down by whatever means possible – the miners had allies within the dominant Labour group on Strathclyde Regional Council, and both rail unions had boycotted the movement of coal during the strike, realising that a defeat for the miners would prove a major setback for trade union bargaining power. In 1985 BR ended the 'closed shop' agreement with the unions to comply with new legislation aimed at weakening the unions. The NUR opposed DOO on the grounds of safety for passengers, but ASLEF ultimately conceded in principle, pending agreement over additional pay and safety issues for DOO drivers, leaving it to the NUR to fight the proposals on behalf of their guard members.

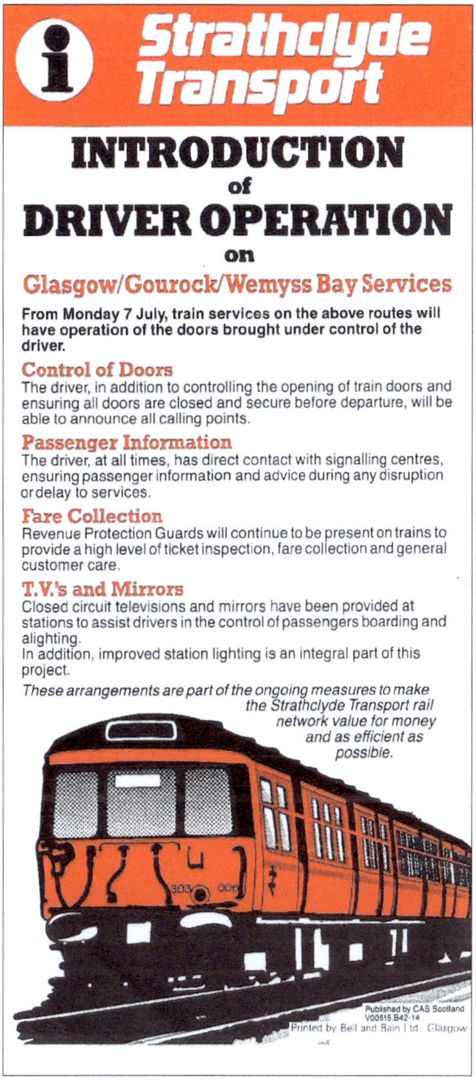

1985 leaflet explaining Driver Only introduction to the public (Darrel Hendrie collection)

Keeping their end of the 1983 review agreement, BR gave notice to the NUR and ASLEF of the introduction in Strathclyde of two DOO door-operating experimental EMUs, starting on Gourock and Wemyss Bay services in 1985, and extending with converted train units to the whole of the Strathclyde network by 1990. Discussions at national and local level failed to resolve the matter and BR stated their intention to commence an experimental DOO with a Gourock train on 2 August 1985, still carrying a guard pending further negotiation at national level, but with doors operated by the driver. Complying with a local NUR decision, the guard refused the duty, was sent home, and 186 of his colleagues walked out on strike. As this was not backed by the NUR nationally, and the strike was unofficial, BR escalated the situation by handing out dismissal notices to all the strikers. It was at this point that Strathclyde Regional Council stepped in, publicly advising BR that they would cease Section 20 subsidies unless the guards were reinstated freeing them to make a local agreement. This was at odds with Strathclyde Region's own ultimatum to the rail unions over the same matter just prior to the strike, reflecting some of the tension within Labour's ranks. Industrial action in opposition to DOO was put to the NUR's guard members nationally, and the majority voted against action. The *Glasgow Herald* of 4 September 1985 reported NUR General Secretary Jimmy Knapp conceding defeat on the issue – paving the way for a 'Strathclyde Agreement' pending further talks at national level. This agreement gave the go-ahead to DOO in the basis of transferring the guards to on-board ticket inspection roles, and monitoring the introduction on the basis that whilst there might be exceptions, there would normally be a driver and ticket inspector on trains.

Chris Green, the then General Manager of BR in Scotland, has said that he believes the unions were broadly onside with tackling the issues in Strathclyde at that time, and ready for difficult negotiations, blaming a minority of local militants within the NUR's ranks for escalating the DOO dispute.[9] An RMT-endorsed history of the union is supportive of the hard line stance taken by local branches in 1985, blaming an unfavourable outcome, involving job loss and other negative impacts, on the low morale within the trade union movement in the wake of the miners' defeat – advantage of which was taken by hard line BR negotiators who themselves escalated the dispute.[10]

Long-term plans for rail development reached some fruition in the lean years of the late 1980s with a short extension of the electrified line from Airdrie out to Drumgelloch. The remainder of this freight line out to Bathgate was dismantled soon afterwards, its use having ended with the area's deindustrialisation, only to be reinstated (as we shall see in the next chapter) in the early years of the next century. Finally in 1989 a new signalling centre was opened at Yoker, eliminating numerous mechanical signal boxes on the North Clyde network over the next few years. In 1990 the non-electrified line was reopened to Paisley Canal.

Bus Deregulation and Privatisation

The pressure to settle the DOO dispute and carry through plans to save Strathclyde's railway network from further contraction and closure was carried out against the background already described, of government disfavour of public services, and desire to let the market determine what people wanted. Bus deregulation had already been announced as part of the Transport Act (1985) and was enacted from October 1986. This was designed to reduce cross-subsidy (profitable services paying for unprofitable ones) and outlaw public subsidy as a means of bringing down fares. It was assumed everyone would prosper as profits were passed onto passengers through reduced fares – questionable assumptions then and now.

Almost all bus services in Scotland at that time were run under the auspices of the state-owned *Scottish Bus Group*, which split into different companies in preparation for deregulation in 1985, with *Kelvin* and *Clydeside* now running services around Glasgow. The former corporation bus service became a company separate from the PTE, *Strathclyde Buses*, and was to be subject to the same level of competition. From the moment of deregulation, buses were chasing passengers and routes, with many small operators setting up services to enter the fray and a few larger ones such as *Stagecoach* and *First Bus* (the formerly public-owned *Grampian Transport*) soon emerging as

The impact of bus de-regulation on the streets of Glasgow 1985, from a later SPTE brochure (author collection)

dominant. Deregulation enabled any private operator to run buses and charge whatever they wished, allowing the larger ones to reduce charges (and in some cases run buses for free) in order to drive out competition. The result initially was chaos – Union Street in Glasgow became jammed with buses (many old redundant London double-deckers came north for the occasion), choking the area with carbon monoxide fumes and endangering health. The Traffic Commissioners looked into concerns expressed by the police and Strathclyde Region but did nothing. This scene was repeated around the country. Transport co-ordination, always problematic, now became very fraught, with SPTE effectively only able to arrange non-commercial bus services: those vital rural and other services requiring subsidy (which would go out to competitive tender), rail services within their area and Dial-a-Bus services for disabled and elderly people. The council, of course, still retained responsibility for the road network and its development and huge sums continued to serve the interests of car users and of course the now private bus companies.

The impact of deregulation on rail services nationally was dramatic: passengers turned to cheaper bus services with an estimated 20–23 per cent loss of usage. This was partially mitigated within a few years as fierce bus competition faded with the emergence of the super bus companies, but in overall terms bus deregulation did the opposite of its stated intentions: higher fares and lower patronage.[11] BR also rose to the challenge by offering competitive fares and heavily promoting HST train services, some on new fast through routes. The opportunity given to the new super bus companies (who had swallowed up competitors not driven out of business) to accrue profit for further investment led several to bid successfully for rail franchises when these became available a decade later: Scottish based Stagecoach became a major player, First Group won the ScotRail franchise for a period, as did former bus company *National Express*; First Group also bought out Strathclyde Buses (privatised three years earlier) in 1996 – the majority of the workforce had purchased £300 shares at the time of privatisation, and now each gained £35,000, a sweetener and boost to the sale that was unfortunately and short-sightedly welcomed at the time by their trade union representatives.

Strathclyde's Ambitious and Tactical Approach to Transport Development

Strathclyde Regional Council, and their SPTE, remained keen to influence transport development as part of their mission to promote an overarching

Launch of the Strathclyde Development Study in 1989 (author collection)

social strategy for their vast geographic area of responsibility in Scotland. Its emphasis was always on serving the disadvantaged as well as the economic interests of the area. This continued through the 1980s with a view to the future. A 1987 review of the 1983 rail plans resulted in a decision to establish the *Strathclyde Transport Development Study*, commencing work in 1988.[12] This review considered that Light Rail Transit (LRT) systems should be examined – despite a submission from chair of the Highways and Transportation Committee, Malcolm Waugh, that this was not feasible in Glasgow as it would cause too much upheaval, and was not working out in Manchester.

The review looked in detail at needs beyond 1990 and, after some background work and preparation, an initial consultation document was issued in April 1989 which considered six options: bus, ScotRail, underground (subway), light metro, light rail, trolley bus, guided bus, 'people movers' (a small scale short distance guided rail system such as located in Birmingham Airport) and water bus.[13] Initial public consultation and visits abroad to look at other urban systems led to some refinement for further consultation and study.

Pressure, it was hoped, could persuade a government who had established constraints on public spending and public ownership to release opportunity and funding. This was laid out explicitly in the next phase – a public consultation document launched in December 1989. Malcolm Waugh stated:

> 'Investment by the Regional Council in public transport and roads is constrained by Government restrictions on capital expenditure and the timing of any investment which is decided upon must be assessed in the context of the limits which apply at any particular time. It is to be hoped that Government constraints do not inhibit timeous implementation of the new roads and public transport projects which are found to be necessary.'[14]

This was both a challenge to the Tory government and warning to readers not to expect immediate change.

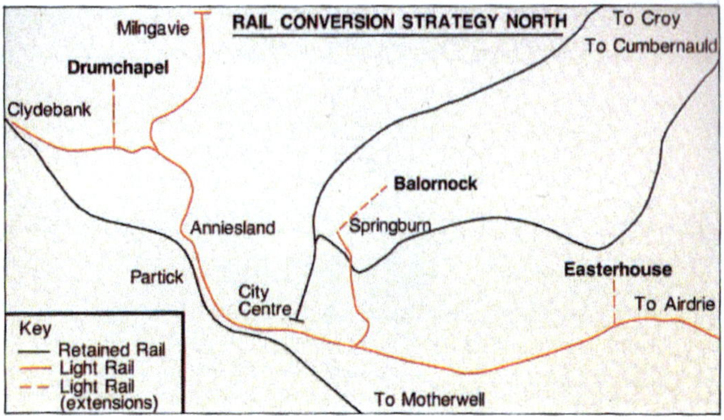

LEFT: From the 1990 SPTE Development Study public consultation document (author collection)

RIGHT: North Glasgow proposals from 1990 (author collection)

The consultation offered the first broad opportunity for public involvement in transport planning in the Glasgow area. The idea of consultation had entered the public arena over the previous decade and was to become an established feature in the years to come. Strathclyde Regional Council became adept at using public opinion: privatisation of Scotland's water industry was successfully resisted after they organised a public referendum on the issue in March 1994. This secured a 1.2 million vote (97 per cent of returned ballot papers on a 70 per cent response rate) against privatisation of water and sewage. Public involvement in transport planning had already been called for by the lobby group Transport 2000 in their 1979 pamphlet *Route Causes*, which outlined how people could get involved using frameworks available at the time through local authorities. This was reminiscent of the wartime ABCA pamphlets in terms of its simple explanation and broad guidance.[15]

Between December 1989 and February 1990, the public in Strathclyde were offered the chance to comment in a set questionnaire on three outline proposals in the transport consultation:

> Strategy A – with an approximate capital cost of £175 million: this was essentially a continuation of existing infrastructure with some bus priority measures, the replacement of trains as required, opening new stations when appropriate, and the development of new suburban rail services on existing lines e.g. a northern Glasgow suburban line between Springburn, Maryhill and Anniesland, and a service between Rutherglen and Coatbridge in Lanarkshire.

Strategy B – with an approximate capital cost of £700 million. This was an ambitious scheme to add to strategy A a new light rail system, some of which was new, and some replacing existing electric suburban trains on the North Clyde and Neilston lines. The new services would be on short, segregated lines out from the City Centre to Drumchapel, Castlemilk, Pollok and Easterhouse – the main peripheral housing schemes with working class populations whose situation Strathclyde Region was set on improving. Finally, the proposal suggested a new rail link between north and south networks using the Tron Rail Bridge.

Strategy C – with an approximate capital cost of £500 million. This involved a light metro service using the lines and replacing South Clyde suburban electric lines to Newton, Neilston and the Cathcart Circle, with short extensions to Castlemilk, Pollok and Newton Mearns ('Southern Metro'). In the north of the city, partially segregated lines would go to Drumchapel, Easterhouse, Balornock, Tollcross and the Riverside ('Northern Metro'). Additionally, a new running loop round city centre streets would join the North Metro at Sauchiehall Street with the South Metro system on Argyle Street. The consultation document advised that this was the preferred option.

43,000 copies of the consultation summary with questionnaire, and 14,000 copies of the full report were distributed. The public were told some facts about present use of services: between 1979 and 1985 bus use had declined by 21 per cent and rail by 9 per cent, but Underground use since modernisation had increased by 13 per cent. Since 1985 bus and rail use had hardly changed although bus mileage had risen by 15 per cent with deregulation. Just less than half of work journeys in the city were by public transport. Neither reducing the rail system further, nor simply leaving things as they were, would prevent increasing car use with its environmental and social consequences, so the report was recommending these options to address pressing issues.

1,540 completed questionnaires were returned along with 125 written responses, eighty of which were from organisations. These were overwhelmingly positive. The majority of responses, 52 per cent of which were from car

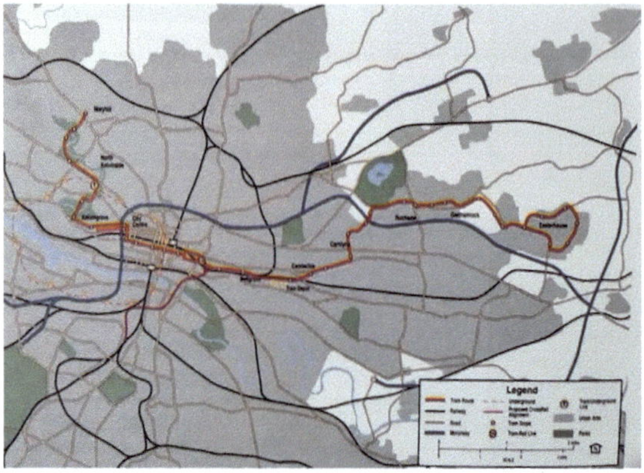

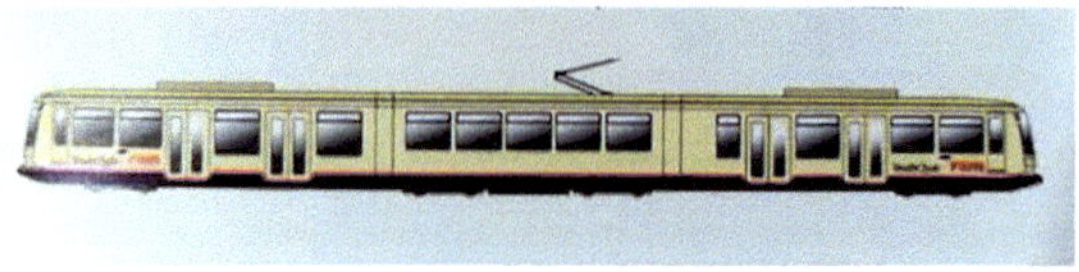

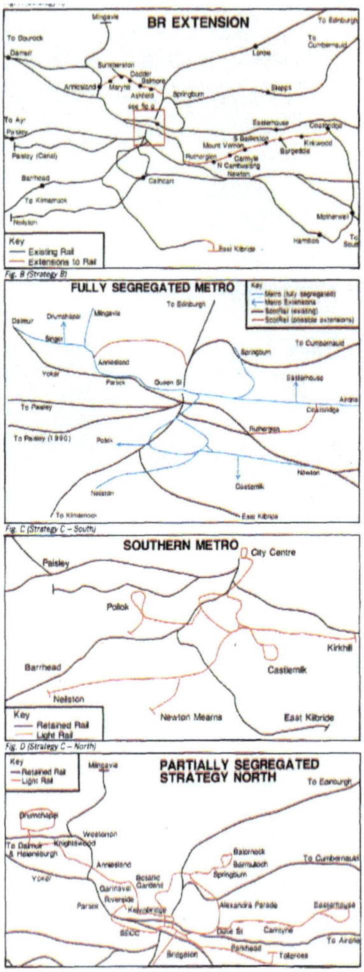

LEFT TOP: Strathclyde Tram proposal 1992 (author collection)

ABOVE: Strathclyde Tram visualisation 1992 (author collection)

RIGHT: Strathclyde Tram line detail from the 1990 public consultation document (author collection)

users, preferred Strategy C, and this became the focus of work over the next few years.[16]

Eventually after further consultations, public exhibitions and promotion, the plans took shape as the £170 million *Strathclyde Tram* formal proposal in 1995. It was hoped that with the public support shown by surveys, and the successful examples in Manchester and Sheffield, the scheme for the 15-mile 'Line 1' from Easterhouse to Maryhill could be in operation by 2000. Other light railway/tram routes would follow, and the main rail network would be improved through a 'Crossrail' link between north and south networks and a link to Glasgow Airport. However, objections were rife – including from Strathclyde buses and the Transport and General Workers' Union who represented their privatised workforce. According to a *Scotsman* report of 27

September 1995, the union argued that this would take away passengers from buses and (perversely given their commitment to the interests of working class people in general), as Line 1 would run through areas of low car ownership, the new service would not reduce car use and traffic congestion in the city.

The formal notice from the SPTE for legislation was published by the Scottish Office on 31 March 1995. Under 60-year-old parliamentary procedure, four commissioners were appointed to look at the proposals and hear objections – four of which had been received, the most prominent being from First Bus who had bought out Strathclyde Buses by this point. The public inquiry convened by the commissioners commenced on 31 March 1996, and on 30 May they announced that the proposal could not go ahead – with no reasons given. Reacting angrily, SPTE's chief executive Steven Lockley stated:

> 'It is so nonsensical that the Parliamentary Commissioners turned it down and we don't know what the reasons were. Strathclyde Buses argued that they could move the same number of people at much less cost. But that is not going to attract people back to public transport. You can't coerce people. You either have to introduce very severe road restrictions which would be electoral suicide for any political party, or make public transport good enough.'[17]

This unhappy ending to eight years and £2 million worth of research and development work came as Strathclyde Regional Council was abolished. Under Tory-led legislation whose clear intention was to end the political power the Region had used successfully to defy government, it was replaced in 1996 by 12 smaller local authorities roughly corresponding to old council boundaries but in many cases divided so as to avoid the enduring Labour Party control many had been subject to under previous systems. The SPTE was renamed *Strathclyde Passenger Transport Authority* (abbreviated to SPT). The statutory change removed powers to franchise rail services and implemented concessionary fares schemes, these later passing to a new government body *Transport Scotland*, the principal role now being planning and co-ordination of transport services in west-central Scotland. Glasgow City Council showed no appetite at the time to fight the Commissioner's decision about the trams, or revamp them, and the moment passed into history.

Rail privatisation and the events described in this chapter negated the possibility of any large-scale developments in the Glasgow area in the 1990s and it would be the next decade before this changed. Whilst public services

Strathclyde Tram artist's visualisation in the final proposal 1995 (author collection)

in general declined during those years, impetus was to change at the end of the decade through devolution and the formation in 1999 of the Scottish Parliament, who would have responsibility for transport matters under the terms of settlement. From now on, a succession of administrations more amenable to public sector infrastructure development held power, albeit with budgets largely determined in Westminster. This will be the subject of the next and final chapter of this book.

Chapter 8

INTO THE 21ST CENTURY

'INTEGRATED RAIL NETWORK: Place rail travel at the core of an integrated transport system, by subsidising rail infrastructure to make it more affordable and resilient than air travel, particularly for mainland journeys in the UK.

PUBLIC TRANSPORT CHEAPER OR FREE: Make public transport cheaper, or free, by reviewing tender processes to focus government subsidies into nationalised public/private partnerships or not for profit transport providers.'[1]

Introduction

The points above were amongst the 81 recommendations of Scotland's *Climate Change Assembly* in 2021, an attempt by Scottish Government to include the participation of citizens in its approach to climate change and its aspiration to meet net-zero emission targets in line with international COP agreements. Early in this book, the contention was laid out that post-war Britain was largely constructed from the aspirations and hopes that drove people towards victory in the Second World War. During the wartime period there was discussion about participation by ordinary people in post-war reconstruction that went beyond normal parliamentary democracy, important though that was and remains. The previous chapters of the book have demonstrated through examination of railway development that people have not generally been consulted until recent times, and, when they have been, they are in favour of quite radical solutions. Instead of consultation, decisions have been made by planners and politicians through trial-and-error methods, and dependence on whether funds were available and not required to meet other priorities. Rising car ownership has lent preference over the years to road infrastructure improvement: with private car and freight carrying vehicle ownership, the user is responsible for the upkeep and fuelling of their asset, so the provision of roads becomes a cheaper option for public sector planners wanting to avoid the capital and recurring expense of trains, trams and light railways – and their necessary infrastructure.

Has this changed in the last quarter century? Certainly, as quoted throughout previous chapters by railway people from both management and trade union backgrounds, road investment has come before railways, creating almost a fait accompli that reached a high with the Beeching cuts of the early 1960s. How could people be attracted back to railways when they were no longer there, or when they provided such poor services? The truth of this is demonstrated by the story of recent rail development in Scotland.

Passenger train use for those who can afford it has maintained a high level having dropped drastically during the Covid restriction years between 2020 and 2022, climbing back up to 84.7 million passenger journeys in 2025, from 63.7 million in 2023. In 1960, the full year prior to Blue Train initial introduction, BR Scottish Region recorded 64.9 million journeys.[3] Given that the recent figures are for ScotRail only and do not include those on cross border and services by other operators, this represents a complete turnaround.

Against a background of social change and faltering progress, this final chapter will cover developments in the first quarter of the present century, and their background in Scotland's devolved parliament and its commitment to a net-zero carbon-free railway system by 2035. Progress to reach this goal has been rapid and ended the association between the railway system and coal: the final journey by a BR steam engine in regular service in Scotland was in 1967, the last deep coal mine in Scotland (Longannet in Fife) closed in 2002 and almost certainly the last opencast operation (House of Water in Ayrshire) in 2020. Britain's final coal-carrying train ran in July 2024 to Ratcliffe-on-Soar power station in Nottinghamshire just before its closure as the last coal-fired power station, marking the end of the long link between rail and coal. The rapidity of these events creates challenges for the future that go beyond climate change imperatives to cease fossil fuel use: Scotland's industrial base across its central belt and the communities that served it were based around coal and its availability for other industrial processes.

The demise of coal and the onset of de-industrialisation has left many central Scotland communities high and dry, so the need to improve their

LEFT: The last BR steam train, an enthusiast special, 1968
(National Archives: Wikimedia Creative Commons)
RIGHT: One of the last coal-hauling trains Didcot Oxfordshire 2013
(Steve Daniels: Wikimedia Creative Commons)

connectivity has to be a factor in the drive for improved net-zero transport networks. Overall planning for the City of Glasgow and its environs, a feature of the Bruce and Abercrombie wartime era reports, was overtaken largely by social and economic change with their long-lasting and often negative consequences.

Interestingly just as the coal miners of Britain, through their comradeship and union organisation, led working-class organisation in Britain over most of the last century, the main transport union, the RMT, are now probably the best organised and most effective trade union in an era of general decline in membership.

The background to 21st-century developments in the Greater Glasgow region also has to be set against the disastrous consequences and lasting impact of the 1986 bus deregulation. Not only did this, as described in the last chapter, raise fares, reduce overall bus availability and obstruct integration between public transport modes, it also created a powerful lobby of very large private bus companies who were set to resist attempts to mitigate or roll back these negative consequences.

Planning the New Glasgow

At the turn of the century and the close of the Victorian era, Glasgow was highly industrialised with a population density, based on rapid tenement building, that was unique in Britain. Alongside such development came endemic poverty. Those who benefited from working-class labour lived outside this morass, hence the suburban railway building described in Chapter 1 which enabled them to travel to work and back home again to fresher air in the leafy suburbs. The city centre where the middle classes worked, shopped and dined out reflected wealth and opulence, in contrast to the teeming tenements that surrounded it. By the 1950s the industrial and shipping trade-based economy of the city was changing for ever – the result of global competition for markets in which Glasgow no longer enjoyed primacy due to Empire and dominance, and changes in social conditions arising from reductions in inequality that accelerated in the post-war period. In Chapters 2 and 3 we saw the city planners' response to this: new towns in central Scotland, peripheral housing schemes in Glasgow that became beset with their own social problems and opportunist building of high-rise flats, many of which were based on design concepts that did not fit the city's damp climate and long winters. With the loss of work opportunity and the dispersal of many of its people, the city's population dropped sharply: from 1,120,000 in 1945, to 650,000 in 2024, although increasing again in recent years. The Greater Glasgow area, including all the urban centres around the city, now has a population of 1.7 million. The city centre itself has changed from a shopping and business focus around its former commercial port and industrial status, to one based on services and tourism. Hotels that seem to increase in number every year feature in the city centre and hospitality is now a major employer. Also numerous are call centres – where online business is managed, and customer interface maintained. Numerous single-person accommodation blocks catering to the universities' foreign students, point to the business ethos of the city's universities.

We have also seen in earlier chapters that there were assumptions about the need for improved road building as railways went into decline that were not based on reality in the west of Scotland. Car ownership across Scotland has indeed seen a steady rise – doubling between 1975 and 2014 to 2,821,000 (including 2,370,000 private cars), more than one for every adult person in the country. In the Beeching years it was assumed that such trends spelt doom for the railways. In Glasgow, however, car ownership remains relatively low: the *Understanding Glasgow Project* analysis of data shows that in 2018 46 per cent of households had no access to a car, compared to 28.6 per cent average across Scotland.[2] The need for public transport in the city remains high, and car use to the centre is generally considered impractical due to congestion, emission and parking availability issues.

Ninety-six million passenger journeys were made on ScotRail trains in 2014–15 (a figure maintained up until the onset of Covid in 2020), proving, just as happened long ago when the Blue Trains were introduced, that people will use railways if they are comfortable, fast and convenient. With continuing rail improvements described below, the figures are even more illustrative: the 2024 official figures show 3.13 million vehicles, 82 per cent of which are private cars with 71 per cent of adults having access to a car, but rail use continues to rise.

In Glasgow, research has shown that some car ownership is forced onto people who cannot afford their use, simply because other means of transport are not available. Unaffordable car ownership is especially prevalent in the large peripheral housing schemes developed in the post-war period where people need to use them to travel to work – many of the lowest paid jobs involve round-the-clock shift patterns, or twilight hour or night working. In some areas cars are the only practical and convenient way to access shops for day-to-day needs. The *Minimum Income Standard* which is updated regularly and used by research bodies and Government to measure poverty and establish benefit levels, now includes car ownership as a necessity for families with children.[4]

ScotRail and Privatisation

Whilst some level of progress had been achieved since the huge leap forward of the Blue Trains in the 1960s, privatisation and its immediate aftermath put a halt to rail development in Scotland in the 1990s. The only major electrification project, planned earlier by BR, was the East Coast Main Line completion which reached Edinburgh in 1991, enabling fast travel between Edinburgh and King's Cross, London. The construction teams also electrified the short branch to North Berwick, and joined the East and West Coast routes by electrifying the 28-mile line from Carstairs to Edinburgh, allowing through passage of electric trains between Scotland's two main cities. It was to be another 14 years before the next initiative which emerged after the 1999 devolution of government.

The mid-1990s privatisation which involved the severe fragmentation described in the last chapter resulted in administrative changes to the railways that affected Scotland. Whilst the ScotRail franchise involved a single tender, trains now ran into Scotland from several different operating companies: LNER, West Coast Main Line (which went through several name changes), Cross-Country, and later Trans-Pennine and most recently LUMO. The two main principal cross-border train stations, Glasgow Central and Edinburgh Waverley were run by the single UK track operator, Railtrack, which, as related in the last chapter, was re-nationalised as Network Rail in the wake of evidence of systematic failure and loss of life in train accidents. The first three years of privatisation saw a gap in investment as profits were given priority – all 27 UK operating companies taking advantage of the successful operations they had taken over from BR, to the detriment of ongoing railway development.

Franchising arrangements changed at UK level in the years following privatisation, giving the devolved Scottish Executive (name changed to Scottish Government from 2005) some powers from 2000, and more authority alongside the national franchising authority from 2005. The ScotRail franchise went for its first seven years (1997 to 2004) to *National Express*, a bus operator who were consequently forced to sell off their rival (to rail) City Link buses. This first franchising operation rejected a management buy-out by ScotRail chiefs. The second franchise was won by another bus operator, *First Group*, who operated the system until 2015. *Abellio* (the international name for the state-owned Netherlands Railways) won the franchise in 2015 over other bidders, apparently by a narrow margin but on the basis of quality over projected subsidy requirement which was actually higher than the bid from National Express.[5]

The Scottish Government formed *Transport Scotland* in 2005 to co-ordinate their responsibilities in all aspects of public transport including ScotRail, but with no direct influence over cross-border services by other companies (except the *Caledonian Sleeper* service which was based in Scotland). Whilst Transport Scotland was given wide enabling powers of scrutiny and investment possibility that would prove positive almost immediately, the Strathclyde Passenger Transport Authority was changed to *Strathclyde Partnership for Transport* (SPT), losing its financial and operational responsibilities for railways, with similar bodies established in other regions of Scotland. *Transport Focus Scotland* emerged as the independent watchdog for rail users, but with no specific powers.

ScotRail Developments 2005 to 2025

The recent historical survey of railway electrification in the UK by experts Buxton and Heath is complimentary in its description of post-privatisation developments in Scotland. It considers that the country has benefitted generally from the commitment by the Scottish Government to an affordable, socially beneficial and environmentally friendly rail network. Electrification has been central to this purpose, and investment has been provided to enable steady development using the accumulated experience and skills of Network Rail staff, Transport Scotland, private contractors and other stakeholders. The authors illustrate this by citing the fact that costs in Scotland for electrification are generally half that of the Great Western mainline project south of the border.[6]

The 1997 Kyoto international agreement on reducing greenhouse gas emissions which came into force in 2005 was embraced by Scottish Government and led to legislation in 2009, updated in 2019, with ambitious targets set that would reduce dependency on fossil fuels and aim for early de-carbonisation of the rail network. Carbon reduction was written into ScotRail franchise contracts and drove investment through Transport Scotland into electrification projects. This built upon plans stemming back to BR days for further Glasgow area rail development and electrification.

The first such project involved the line between Hamilton and Larkhall in 2005 – a four-mile branch that had been closed completely 40 years earlier, marking the first reopening of a closed Scottish line in 25 years. This extended the 1979 Argyle Line development and enabled the running of electric trains through from Dalmuir to Larkhall. Another aspect of this investment involved a new spur between Maryhill and Anniesland which allowed introduction of a DMU suburban service from Queen Street through the north of Glasgow.

The team that had constructed the Larkhall extension then moved on to upgrade sections of the West Coast Main Line (WCML) which was completed in 2008. They then commenced a major project – the reopening and electrification of the line between Airdrie and Bathgate, a proposal first looked at in 2002, and funding agreed by the Scottish Executive in 2003. This involved a £300 million investment to reopen 14 miles of railway closed completely with the removal of tracks in 1983. The section of the route from Bathgate on to Edinburgh, which had been retained as a freight line, had reopened for passengers in 1986. This ambitious project involved extensive work reconstructing the track to modern standards with proper drainage over the wetlands traversed, building bridges to replace level crossings and effectively creating a new-build railway rather than simply reinstating the old one. Some structures including houses in Plains had to be demolished, and new stations constructed with park-and-ride facilities. Environmental and ecological issues had to be addressed including working around the presence of protected species such as badgers, otters and crested newts. Much of the abandoned line west of Bathgate had been turned into a walking and cycling track linking Glasgow and Edinburgh, so this had to be reinstated alongside the double-track railway. Planning work involving 150 staff based across Britain commenced in 2007 and the physical work started in 2009, the two main contractors being Carillion PLC for the civil engineering and Balfour Beatty for the permanent way and electrification. Despite the best efforts of the weather – a poor 2009/10 winter with very heavy rainfall – the line was completed on time and on budget in December 2010. In total the project involved 4.5 million man-hours of work which would be justified by an expected 12,000 passenger journeys per day.[7] The very detailed official 2013 evaluation suggests that patronage increased overall after opening and that a number of objectives had been advanced if not met: these included broad social aims such as the promotion of social inclusion within the deprived communities along the route, and reduction of congestion on the parallel Falkirk line. Whether it had reduced car use on the M8 motorway between the cities could not be effectively measured.[8]

The next electrification project concerned the eight-mile route from Glasgow Central Station to Paisley Canal which had been reopened to DMU passenger trains in 1990. Almost half was already in place so this only involved some five miles creating capacity for the overhead wires whilst existing services continued. Although work only took place at night and at weekends, the route was ready after 44 workdays, opening in December 2012. Of note is the fact that electric services were managed without the necessity for new stock, and that the DMUs released were diverted to the new Waverley line from Edinburgh (of which more below).

A similar project to the Paisley Canal improvement involved the 16-mile Rutherglen to Coatbridge route, first mooted for electrification in 2006, and finally given Scottish Government approval in 2013. The purpose was to improve the overall suburban network both from Glasgow out to Lanarkshire but also between its major conurbations – this author recalls working in the area in the 1970s when travel between towns in Lanarkshire was complex because all the main road and rail routes went straight to Glasgow. This line too had been reopened to DMU passenger use in 1993, but only as far as Whifflet on the outskirts of the large town of Coatbridge. The electrification involved extending the service to a reopened station at Coatbridge Central. Whilst the work was not completed in time, as projected, for the 2014 Glasgow Commonwealth Games, it was in service by the end of September 2014.

Early in the life of the Scottish Parliament, the then Labour-Liberal Democrat coalition determined a bold plan to modernise the main routes across central Scotland. This would involve the electrification of the main line from Glasgow to Edinburgh, the line from there to Stirling, and the loop through Falkirk Grahamston Station. Announced in 2006, and agreed to also by the successor Scottish National Party (SNP) government, the project took the title *Edinburgh Glasgow Improvement Programme* (EGIP). Revised in ensuing years it included short extensions from Stirling to Dunblane, and the line to Alloa which had been reopened for passengers in 2008. It also involved electrification of the route through Cumbernauld from Glasgow Queen Street which joined the Falkirk Grahamston and Stirling lines at Greenhill, the purpose being to provide a useful diversionary route as well as an upgrade for suburban services to Cumbernauld. As envisaged from the outset, the main line at the centre of the project would offer six trains an hour between Edinburgh and Glasgow, a 42-minute journey time for the fastest trains, and an increase in passenger use of 30 per cent. In all the project took in 126 route miles in an area that includes two thirds of Scotland's population.

The proposals became a reality with an announcement by Transport Scotland in July 2012: a rolling programme of work that would cost £650 million for the first phase that did not include the Stirling, Dunblane and Alloa extensions. As with other projects this too involved engineering challenges, the raising of bridges, lowering of track bed and considerable tunnel work. The latter included a problematic tunnel west of Edinburgh at Winchburgh; the initial plan suggested this should be by-passed completely with a new two-mile surface line – the 'Almond Loop' but this idea was abandoned on cost grounds. Ultimately a team of Austrian engineers worked day and night in the dank tunnel, only closing it to services for six weeks in the summer of 2015 and completing work at a cost of £17 million. The work on the tunnel just outside Glasgow Queen Street Station involved the closure of the High Level station for 20 weeks and diversion of services to the Lower Level and to Glasgow Central – making full use of the previously electrified and fast route to Edinburgh through Bathgate. Other developments associated with the project included a new station at Edinburgh Park, and three completely rebuilt stations at Edinburgh Haymarket, Cumbernauld and Glasgow Queen Street. By 2016 funding had been found for the Stirling, Alloa and Dunblane extensions, and also for the 46-mile line between Edinburgh and Glasgow via Shotts. The latter meant that the four central belt connections between

Edinburgh and Glasgow were now all to be electrified, bringing populated and less populated areas between them fast and regular services that had not been seen for many years. In total, EGIP electrification between 2014 and 2019 had been brought to 202 route miles, increasing Scotland's electrified lines to over two-thirds of the total network.[9]

The EGIP lines and station redevelopments were opened in stages as the construction teams moved from location to location: the main Edinburgh–Glasgow route via Falkirk High fully opened for electric services in December 2019, shortly before the Covid pandemic delayed continued development. For this and other services, 46 new three-car and 24 four-car '385' train units had been ordered from Hitachi, the Japanese manufacturer who had built an assembly plant in Newton-Aycliffe, County Durham. The ScotRail express units included a limited number of first-class compartments – the only EMUs in Scotland to do so. Their introduction was delayed by issues with the shape of the drivers' windscreens obscuring vision with signals at night, so they had to be replaced; the units are not universally popular with crews due to the lack of space in the cabs and the restricted forward vision due to the position of the gangway.

The Glasgow Queen Street redevelopment was completed in October 2021 at a cost of £120 million, bringing improved visual impact from the

Class 385 EMUs Edinburgh Waverley Station (E235JREMU: Wikimedia Commons)

city's George Square (the fine Victorian glass arch station had been obscured by post-war construction) and lengthened platforms to accommodate the four-trains-an-hour express services to Edinburgh.[10]

Much of this progress has reversed the Beeching cuts, and demonstrated how short-sighted they were in terms of the later expense of reviving long-closed lines and their infrastructure, and utilising parallel alternative mainline routes to maintain services flexibly. Whilst outside the scope of this book, this is nowhere better illustrated than by the partial reopening of the Waverley Route that ran until closure in the late 1960s from Edinburgh 90 miles across the south of Scotland to Carlisle. This left a vast area of Scotland bereft of accessible railways and the campaign to reopen never ceased from the moment of closure. However, it was not until the 1990s that any effort was made to save the former track bed from redevelopment so that there was a possibility of future reopening. In 2015 the 30-mile section out to Tweedbank from Edinburgh was reopened for DMU services, and continuation to Carlisle remains under consideration.[11]

Electrification in central Scotland continues: in 2024 the line to Barrhead from Glasgow Central was fully electrified and at the end of 2025 the 13-mile route went out to a newly built station at East Kilbride. Both of these had been recommended for electrification in the 1951 Inglis Report, so this had taken a long time to reach fruition. This work almost completes the electrification of existing routes in the immediate Glasgow area. On the east coast, work is underway to bring electric services to routes into Fife, and the reopened line to Tweedbank.

The progress described represents a remarkable achievement during a period when public services have been struggling to keep up with growing need in other areas including health and social services as well as other infrastructure requiring large capital investment. The 2021 Scottish Government Investment Programme for the period 2021/22 to 2025/26 for rail improvement would involve £1.2 billion additional monies including £550 million for de-carbonisation of the rail infrastructure. This was on top of the £3.8 billion for continued operation and routine sustainable renewal of equipment on the rail system. For the same period roads would get an additional £1.5 billion for improvement, including £60 million for climate adaption and resilience in the trunk road network and £120 million for introducing electric buses. Much of the new roads investment was for specifically named projects (for example, A9 carriageway duelling). Taking 2023/24 during this period, normal roads maintenance expenditure for motorways and trunk roads cost £374.6

million, and smaller roads maintenance by local authorities cost £453 million (55 per cent of their total transport budgets).[12] Whilst detailed analysis is a complex matter and beyond the scope of this book, it does seem that rail investment by the 2020s was outstripping that of roads.

The worldwide drive towards net zero through agreement made at the regular COP conferences has driven government policies at UK and Scottish levels. The Scottish Government's target for rail de-carbonisation through the removal of DMUs was initially to be achieved by 2035, but in 2025 deferment to 2045 was announced as a realistic target for necessary fleet replacement.[13] At this time the ScotRail fleet included 203 EMUs and 145 DMUs, and their future replacement involving new technologies is described below.

What Happened to Transport Co-ordination?

So far, this chapter has painted a positive picture of 21st-century developments in Scotland. Sadly, the damage inflicted by the bus de-regulation that took place 40 years ago, has yet to be undone, and this remains a barrier to progress. In England and Wales, where rail electrification has not proceeded at the same pace, legislation has improved the situation, and examples such as Manchester's *Bee Network* showcase co-ordination that meet aspirations for active travel, rapid urban transit and sustainable net-zero transport services.[14] In Scotland the regional transport authorities established by the Transport (Scotland) Act 2005 lost the rail franchising powers of their predecessor authorities (SPTE and others), their main role now being to maintain infrastructure (bus stops and bus stations), plan service provision and investment and directly run some subsidised services and dial-a-bus schemes for disabled travellers (and in Strathclyde, the Glasgow Subway). Their insufficient budgets since have stifled development. Meanwhile the private bus companies have reaped the rewards of Scottish Government established concessionary travel schemes, setting high single fares that are off-putting to paying travellers, but highly profitable as there are no limits for concessionary card holders travelling by bus within Scotland. In Scotland in 2023 the Government invested £470 million in the deregulated bus system – more than half through the Concessionary Card Scheme.

The Transport (Scotland) Act 2019 was designed to aid the process of re-regulating bus companies by giving Strathclyde Partnership for Transport, and the other authorities, powers to franchise services – and bring them back into public ownership. Similar legislation in England and Wales sped up such processes (as in Greater Manchester) but in Scotland this has not happened. Not only does SPT lack the necessary funds to proceed but the legislation creates an obstacle in requiring the approval of an independent panel convened by the Travel Commissioner, before plans can be progressed. SPT agreed to bus re-regulation at the start of 2025, and the Scottish Government budget for 2026/27 includes an investment of £4 million to enable the creation of a business case to create the conditions for public ownership of buses. In the view of the campaigning group, *Get Glasgow Moving*, this is pivotal to public transport development and integration in the region – including the *Clyde Metro* proposals discussed below. Every delay pushes back climate change net-zero targets, particularly the goal of reducing private car use.[15]

Infrastructure changes and proposals involving public spending now have to go out for public consultation, a feature that did not exist until this century, and although there might be cynicism about the extent to which this influences eventual decisions, it does ensure opportunity.

21st-Century Transport Plans for Glasgow

Emasculated in the 1980s and lacking power and authority since, *Strathclyde Partnership for Transport* (SPT), the successor to SPTE created under the Transport Scotland Act 2005, has struggled since to realise the role it should be playing in developing a modern integrated system in the West of Scotland and the City of Glasgow. Bus de-regulation has played a key role in their diminution.

In 2005 SPT commissioned a *Long Term Strategy for the Subway*, publishing proposals in 2007. This went back to ideas from previous decades: extending the existing Subway outwards to the east end, south side and north-west of the city. The proposals suggested upgrading the system to standard railway gauge, using the closed line from Exhibition Centre to Maryhill, and some existing suburban lines in the south and east of the city centre, to create an integrated network. However, the total cost envisaged, at £10 billion, quickly kicked it into touch at a time when the economy was reeling from the bank crisis; revitalising the bus network along rapid transit lines was considered a more affordable option.

In the same period the idea of a rail link to Glasgow Airport was again looked at by Transport Scotland and ScotRail. A detailed Glasgow Airport Rail Link (GARL) proposal was completed in 2006 that involved a spur from

Edinburgh Tram on the Newhaven extension 2025 (author)

the existing Inverclyde line west of Paisley Gilmour Street Station which would take passengers directly to the airport from Glasgow Central Station four times an hour. The project was cancelled as a public spending cut in 2009. The GARL concept was revived in 2016 by Glasgow City and Renfrew Councils, this time as a tramway with the title Glasgow Airport Access Project (GAAP). In 2017 SPT commissioned an 'internal briefing note' that revived ideas about extending the Glasgow Subway, which considered the limitations of the existing four-feet gauge for future development – a system incompatible with 21st-century standards of access and comfort and likely the main reason for its sidelining in the years since.[16]

The airport link proposal has since been absorbed into a much wider long-term (30-year) *Clyde Metro* project which at the time of writing is still at early planning stages, but, with government approval and incorporation into Transport Scotland's work plan, represents hope for the future. Clyde Metro proposals involve the development and conversion of existing suburban railway lines into light railway routes with new-build extensions and tram-trains on reopened lines and city streets; the proposals are couched in terms of 'what might be' rather than 'what will happen'. Detailed ideas behind the scheme are available at the time of writing on the SPT website: these are ambitious proposals in the spirit of Bruce and Fitzpayne's 1940s

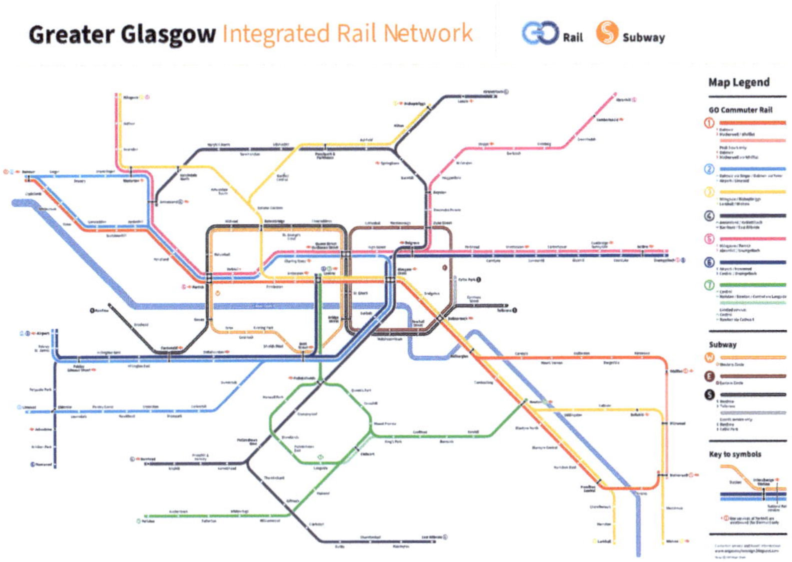

TOP: A visualisation of how the Glasgow rail network might look in the future (Angus Doyle)

ABOVE: SPT Clyde Metro options; the image is illustrative only as development work on Clyde Metro continues with no decision made on specific modes at time of writing (SPT)

visions.[17] First proposed in 1999, trams finally returned to Edinburgh on the other side of Scotland in 2014; these were well over-budget and controversially executed but have become popular in use, and are now extended to cover 11.5 miles from the airport to Newhaven.

What of the Future?

This book has described electrification railway projects and their background social and policy history, touching at times on the politics that drove them both backwards and forwards. At the start of this story, coal was still king but the use of electric energy as a direct power source had started making inroads from the late 19th century in urban tramways, and on the railway system from the 1920s onwards. In the hundred years since then trams disappeared but have enjoyed a revival this century, and railways were reduced in favour of cars and road transportation. Now, railways too are enjoying renewed interest due to the impetus of climate change and all the negative issues caused by rising private car use, especially within and between conurbations. This is barely impacting on freight yet but has certainly done so with passenger transport – railway use increasing as services are improved and sped up. Journeys between London and Scotland by train are able to compete with air travel for speed and convenience. Hesitancy about radical developments like HS2 has arisen mainly for cost reasons, and the truncated result of a new line from London to Birmingham criticised for its relevance and purpose.

Over most of the period described in previous chapters, electrification was seen as key but not cost effective on rail lines that were considered socially useful but unprofitable – many saved from the Beeching Axe including those north of the central belt of Scotland. Now new technologies are under development to tackle such questions and move away from the diesel traction which was originally intended anyway (in early BR days) to be a stop-gap.

Seventy years after an early experiment on the Ballater Branch line, Battery-Electric powered Multiple Units (BEMUs) are being trialled in the UK as technological advance reduces their size and increases their capacity. Hybrid bi-mode trains that can operate on electrified lines, but with diesel motors for other routes, have been around for many years. Taking this idea to a new level, bi-mode units and locomotives are under development that use electrified lines for direct power and also to charge batteries that can then be used on other lines; some are tri-mode, having diesel motors too. The idea of 'discontinuous electrification' is being discussed for the Fife lines and others in Scotland and forms the basis of Transport Scotland's planned replacement of their DMU fleet. Such schemes can avoid locations for electrification that would prove expensive – tunnels and low bridges, and long sections between power sources. This could be the future for the Far North and West Highlands and bring back into play the use of sustainable hydro-electric power sources. The railway network includes land property surrounding the system that could be used to generate electricity, through solar power or hydro-electricity, feeding rail overhead lines almost directly without the need for extensive power lines for inefficient transmission across the countryside.

Class 93 Tri-Mode locomotive in use in the UK, 2025 (Kitmasterbloke: Wikimedia Creative Commons)

Hydrogen power is also under development: an old shunting engine was converted from diesel to hydrogen in 2021 on the Severn Valley Heritage Railway (SVR) in Shropshire. More recently the technology has been applied to a redundant EMU on the SVR to create the HydroFLEX train which has been tested on main lines at up to 90 mph. Hydrogen production is centralised, and its domestic market is growing in the UK – with a plant on an old coal-fired power station site, HyMarnham in Nottinghamshire, that is rail connected. Its owners are currently considering using rail for distribution. If this technology takes off, in the future perhaps hydrogen could replace coal as the mainstay of railway freight business, as well as a source of motive power.[18]

Sixty-five years after the Blue Trains were introduced to the Glasgow region, it does seem that the vision of their architects has won out over those who preferred to invest in modes of transport that have proved unsustainable. The ghost of Beeching has been finally laid to rest and rail transit has a future that will lend to a reduction of inequalities and a carbon-free environment. That was the dream of those post-war public service-orientated planners whose visions form the core of this book. Whilst public participation in the running as well as ownership of transport services has only taken faltering steps forward since the Second World War, perhaps there is hope here too: in the second quarter of

HydroFLEX prototype train on the Severn Valley Railway 2025 (Chris Allen: Wikimedia Creative Commons)

the 21st century there is considerable interest by environmentally conscious and well-informed grass roots campaigning groups over transport infrastructure matters. They use social media to share information and concerns, hold public meetings and lobbying events, and circulate petitions and respond to consultations. Protest and political lobbying are not new but in the past were perhaps ineffective in influencing public opinion and voting at elections over matters such as public transport. Groups like *Get Glasgow Moving* will not go away: citizen voices are louder than ever.

APPENDIX

Main and Suburban Electrified Railways in Scotland – Dates of Opening for Service 1960–2025

Date	Route	Route Mileage Electrified	Accumulated Mileage
November 1960	Helensburgh–Airdrie via Glasgow Queen Street Low Level; Clydebank Loop; Balloch Milngavie & Springburn Branches	49.75	
May 1962	Glasgow Central–Cathcart Circle; Neilston; Motherwell via Kirkhill	28.75	78.50
June 1967	Glasgow Central–Gourock; Wemyss Bay	37.25	115.75
May 1974	Glasgow Central–Hamilton Circle; Wishaw; Lanark branch	24.50	140.25
May 1974	WCML Motherwell to Carlisle	89	229.25
May 1974	Carluke to Lanark	3	232.25
November 1979	Argyle Line: Partick to Rutherglen via Glasgow Central Low Level	5.25	237.50
September 1986	Paisley Gilmour Street to Ayr; Largs and Ardrossan Harbour branches	51.25	288.75
May 1989	Airdrie–Drumgelloch	1.5	290.25
July 1991	ECML Edinburgh–Berwick-on-Tweed	57.5	347.75
July 1991	Carstairs–Edinburgh	28	375.75
July 1991	Edinburgh–North Berwick	5	380.75
December 2005	Larkhall branch	4.25	385

December 2010	Drumgelloch–Edinburgh via Bathgate	32.25	417.5
December 2012	Glasgow Central–Paisley Canal	8	425.5
May 2014	Motherwell–Cumbernauld	16	441.5
September 2014	Rutherglen–Coatbridge	16	457.5
December 2017	Glasgow Queen Street–Edinburgh via Falkirk High – & diversionary lines	94	551.5
December 2018	Stirling, Alloa, Dunblane	32	583.5
April 2019	Glasgow Queen Street-Edinburgh via Shotts	46	629.5
August 2023	Glasgow Central–Barrhead	9	638.5
December 2025	Glasgow Central–East Kilbride	7	645.5

BIBLIOGRAPHY AND ONLINE SOURCES

Primary Sources:

Abercrombie, Sir P. and Matthew, R.H. (1946) *The Clyde Valley Regional Plan* (known as the 'Abercrombie Report') Glasgow, University Press

Beharrell, L.V. (1962) 'Glasgow Blue Trains: Some Implications' *British Transport Review* VI (5) April 1962

Blake, G. (1960) *Glasgow Electric – the story of Scotland's electric railway* British Railways Scotland

BR Public Relations Division (undated but probably 1962) *Internal Relations Case Study – The Glasgow Blue Trains* British Railways

British Rail (1978) *Railway Electrification: A British Railways Board Discussion Paper*

British Rail, Strathclyde Transport (1983) *Review of Strathclyde Rail Network – the Way Ahead*

British Railways Board *Annual Report and Accounts* (various years)

British Railways Scottish Region (1954) *Report of the Regional Committee on the Electrification of Glasgow Suburban Lines*

British Transport Commission *Report and Accounts* (various years)

British Transport Commission (1955) *Modernisation and Re-Equipment of British Railways* London, British Transport Commission (the Modernisation Plan)

British Transport Commission (1960 2nd impression) *Modernisation of British Railways – The System of Electrification for British Railways* London, British Transport Commission

British Transport Review journal (1950–64)

Buckton, R. (1982) *Transport – The Forgotten Social Service – lecture given to the Chartered Institute of Public Transport by Ray Buckton, General Secretary, ASLEF 5th April 1982* London, ASLEF

Corporation of the City of Glasgow First Planning Report (1945) (known as the 'Bruce Report') Corporation of the City of Glasgow Highways and Transport Committee

Corporation of the City of Glasgow, Scottish Development Department (1966) *Greater Glasgow Transportation Study Vol.1: Current Travel Patterns in the City of Glasgow – Origin Destination Survey 1964-65*

Corporation of the City of Glasgow, Scottish Development Department (1968) *Greater Glasgow Transportation Study Vol.2: Forecast and Plan*

Craig, G.B., Logie, J.G., Scott, K.D. and Taylor, M.S. (1980) 'The Transclyde railway network: Kelvinhaugh burrowing junction' *Proceedings of the Institution of Civil Engineers* Part 1 1980, August, pp.369-389

Daniel, A.W.T. and Calvert, R. (November 1969) *Electrify Now – A compelling case for further railway electrification in Great Britain, especially between Crewe and Glasgow* London, National Council for Inland Transport

Davis, A. (November 1959) 'Glasgow Suburban Railway' *Design* (the magazine of the London-based Council of Industrial Design)

Electrification of Railways – Report of a Committee Appointed by the Railway Executive and the London Transport Executive (1951) London, British Transport Commission

Evening Times (Glasgow) (various)

Failures of Electric Multiple Units on British Railways Final Report (1962) London, HMSO

Fitzpayne, E.R.L (1948) *A Report on the Future Development of Passenger Transport in Glasgow* Corporation of the City of Glasgow

Glasgow and District Transport Committee (1951) *Passenger Transport in Glasgow and District* (known as the 'Inglis Report') London, British Transport Commission

Glasgow Herald newspaper (various)

Greater Glasgow Passenger Transport Executive (1974) *Development of CLYDERAIL – Application to Secretary of State for Scotland for Infrastructure Grant* Glasgow, GGPTE

Greater Glasgow Passenger Transport Executive (1979) *Trans-Clyde News No.3*

Hansard (various) House of Commons and House of Lords Record of Proceedings

Highway Statistics (1965) Ministry of Transport, London

Ministry of Transport (1931) *Report of the Committee on Main Line Electrification* (known as the 'Weir Report') London, HMSO

Ministry of Transport (1962) *Railway Accidents – Failures of Multiple-Unit Electric Trains on British Railways – Final Report* London, HMSO
National Union of Railwaymen (1963) *The Misshaping of Britain's Railways* London, NUR
National Union of Railwaymen (1979) *Railway Electrification – Submission to the Secretary of State for Transport from the National Union of Railwaymen* London, NUR
Papers of the British Railways Electrification Conference 1960 (1960) British Railways
Public Transport in Strathclyde Region – The Way Ahead – Issues and Options for Assessment (April 1989) Glasgow, Strathclyde Regional Council
Public Transport for the 21st Century – Strathclyde Transport Development Study Consultation Document (December 1989) Glasgow, Strathclyde Regional Council
RAIL fortnightly magazine (various)
Railway Magazine (various)
Report by Working Group on Supported Rail Services (September 1987) Glasgow, Strathclyde Regional Council
Report of the General Manager Year 1st June 1955 to 31st May 1956 (1956) Glasgow Corporation Transport Committee
Royal Commission on Transport (1930) *Final Report – The Co-ordination and Development of Transport* London, HMSO
Savage, C.I. (1957) *Inland Transport* London, HMSO
Scotland on Sunday newspaper (various)
Scotsman newspaper (various)
Scottish Government website (various – details listed at end of Chapter 8)
The Railway Year Book 1922 London, Railway Publishing Company
Trains Illustrated (retitled *Modern Railways* in January 1962) (various)
Transport Age magazine (1957–65)
Transport Scotland website (various – details listed at end of Chapter 8)
Understanding Glasgow website (2026)
Walker, P.J. (1974) 'Light rapid transport for Glasgow' *Modern Tramway and Rapid Transit Review* June 1974
Wilson, J.S. (1990) 'Strathclyde Studies Light Rail' *Modern Tramway* September 1990

Secondary Sources:

Abbis, J. and Lumsdon, L. (1979) *Route Causes – A Guide to Public Transport Plans* London, Transport 2000/National Council of Social Service
Abrams, L., Kearns, A., Hazley, B. and Wright, V. (2020) *Glasgow – High Rise Homes, Estates and Communities in the Post-War Period* London, Routledge
Army Bureau of Current Affairs (September 1942) *Town Planning ABCA No.27* London, War Office
Army Bureau of Current Affairs (November 1943) *Basic ABCA Bulletin* London. War Office
Bagwell, W.S. (1963) *The Railwaymen – The History of the National Union of Railwaymen* London, George Allen & Unwin; Gordon. A. (2012)
Baker, A. (1989) *The Cairo Parliament 1943–44 – An experiment in military democracy* Leigh on Sea, Partizan Press
Berlin, M. (2006) *Never on Our Knees – A History of the RMT, 1979-2006* London, Pluto
Beveridge, W. (1942) *Social Insurance and Allied Services Cmd 6404* London, HMSO
Bonavia, M.R. (1981) *British Rail – The First 25 Years* Newton Abbot, David & Charles
Bond, R.C. (1975) *A Lifetime with Locomotives* Cambridge, Goose & Son
British Railways Press Office (1943) *Facts About British Railways in Wartime* London
British Railways Press Office (1944) *British Railways in Peace and War* London
British Railways Press Office (1945) *It Can Now Be Revealed – More About British Railways in Peace and War* London
British Railways *Your British Railways 1951*
Buxton, J. and Heath, D. (2025) *Lines of Power – The Electrification of British Railways – Trials Tribulations and Triumphs* Lewes, Unicorn
Calder, A. (1969) *The People's War – Britain 1939–1945* London, Jonathan Cape
Campbell, R.W. (1919) *Snooker Tam of the Cathcart Railway* Edinburgh, W.&R. Chambers
Clough, D. (2014) *The Modernisation Plan – British Railways' Blueprint for the Future* London, Ian Allan
Communist Party (1943) *How to End the Muddle on the Railways* London
Communist Party (1944) *Transport for the People: Proposals for Post-War Organisation of British Transport Services* London, Communist Party
Cousins, J. (1986) *British Rail Design* Copenhagen, Danish Design Council
Crump, N. (1947) *By Rail to Victory – The Story of the LNER in Wartime* London, London & North Eastern Railway

Curl, A., Julie Clark, J. and Kearns, A. (2018) 'Household car adoption and financial distress in deprived urban communities: A case of forced car ownership?' *Transport Policy* 65, pp. 61-71
Darwin, B. (1946) *War on the Line – The Southern Railway in Wartime* London, Southern Railway
Deans, B.T. 'Glasgow Subway in the 1920s' *Scottish Transport No.33 Special Transclyde Issue* (December 1979)
Dennis, G. (2024) *How the Railways Will Fix the Future* London, Repeater
Faulkner, R. and Austin, C. (2023) *Signals Passed at Danger – Railway Power and Politics in Britain* Manchester, Crecy Publishing
Festival of Britain Guide (1951)
Florian, U. (2010) *Modernising Glasgow – Tower Blocks, Motorways and New Towns 1940–2010* Glasgow School of Art online research repository https://radar.gsa.ac.uk/5782/
Fowler, J., Edwards, R. and Wilson, J. (2025) 'Britain's First Net Zero: Turning the Lights On and Railways Off 1953-73' *Enterprise & Society* (2025): 1-29
Get Glasgow Moving https://www.getglasgowmoving.org/
Glen, A. (2010) *Airdrie–Bathgate Rail Link – Reconnecting Communities* Isle of Man, Lily
Glen, A. (2018) *Transforming the Railways of Central Scotland* Isle of Man, Lily
Glen, A. (2020) *Glasgow Queen Street – A Railway Station Renaissance* Isle of Man, Lily
Gordon, A. (2012) *Unity is Strength – The National Union of Railwaymen – 100 Years of Industrial Unionism* London, RMT
Gourvish, T.R. (1986) *British Railways 1948–73 – A Business History* Cambridge, Cambridge University Press
Haresnape, B. (1983) *British Railways 1948–83 – A Journey by Design* London, Ian Allan
Hilton, W.S. (1986) *The Plug Dropper* London, Trade Press Publications
Howat, C.J. (2023) *Glasgow Blue Trains – Class 303 and Class 311 EMUs* Stroud, Amberley
Hutt, A. (1937) *The Post-War History of the British Working Class* London, Gollancz
Johnston, T. (1920) *The History of the Working Classes in Scotland* Glasgow, Forward Publishing p358-359
Johnston, T. (1943) *Report of the Committee on Hydro-Electric Development in Scotland Cmd 6406* Edinburgh HMSO
Kernahan, J. (1980) *The Cathcart Circle* Falkirk, Scottish Railway Preservation Society
Kernahan, J. (2011, 2nd ed.) *The Cathcart Circle* Lydney, Lightmoor Press
Kisch, R. (1985) *The Days of the Good Soldiers – Communists in the Armed Forces WWII* London, Journeyman
Kynaston, D. (2007) *Austerity Britain 1945–51* London, Bloomsbury
Labour Manifesto 1945: *Let Us Face the Future – A Declaration of Labour Policy for the Consideration of the Nation* http://www.labour-party.org.uk/manifestos/1945/1945-labour-manifesto.shtml
Lamont, A. (1945) *How Scotland Lost Her Railways – and the case for Nationalisation on a Scottish basis* Glasgow, The Scottish Secretariat (Scottish Nationalist Party)
Little, S.M. 'Greater Glasgow's Railway Network' *Scottish Transport No.33 Special Transclyde Issue* (December 1979)
Mackenzie, S.P. (1992) *Politics and Military Morale – Current-Affairs and Citizenship Education in the British Army 1914–1950* Oxford, Clarendon Press
McKinnon, J. (1921) *The Social and Industrial History of Scotland: From the Union to the Present Time* London, Longman
Macleod, A.B. (1948) *The McIntosh Locomotives of the Caledonian Railway* London, Ian Allan
Malm, A. (2016) *Fossil Capital – The Rise of Steam Power and the Roots of Global Warming* London, Verso
Meighan, M. (2024) *Glasgow – A Transport History* Stroud, Amberley
Millar, A. (1985) *British PTEs: 1 – Strathclyde* London, Ian Allan
Millions Like Us – Report of the Conference of Engineering and Shipbuilding Workers called by the Shop Stewards national Council at the Stoll Theatre London, March 12th 1944, London
Mitchell, A. (1995) *Election '45 – Reflections on the Revolution in Britain* London, Bellow Publishing
Mullay, A. (2006) *Scottish Region, A History 1948–1973* Stroud, Tempus
Murray-Smith, D. *Aberdeen to Ballater by BMU – Notes on the Battery Railcar Experiment* https://www.railscot.co.uk/articles/Aberdeen_to_Ballater_by_BMU:_Notes_on_the_Battery_Railcar_Experiment/
Nash, G. (1946) *The LMS at War* London, London Midland & Scottish Railway
Noble, T. (1990) *Exploring Strathclyde by Rail* North Yorkshire, Leading Edge
Nock, O.S. (1973) *The Caledonian Railway* London, Ian Allan
Nock, O.S. (1974) *Electric – Euston to Glasgow* London, Ian Allan
North British Railway (1894) *Mountain Moor and Loch* London, Sir Joseph Causton & Sons
Oakley, C.A. *The Last Tram* (1962) Glasgow, Corporation of the City of Glasgow
Owen, J. (1943) *Spotlight on the Clyde* Glasgow, Communist Party Scottish Committee
Paterson, A.J. (1969) *The Golden Years of the Clyde Steamers (1889–1914)* Newton Abbot, David & Charles
Patten, B. (1994) *Glasgow in the Age of the Tram, 1950–62* Kettering, Silver Link
Robertson, D.S. (1998) 'Pulling in opposite directions: the failure of post-war planning to regenerate Glasgow' *Planning Perspectives* 13.1 pp53-67

Ross, D. (2018) *Scottish Railways 1923–2016 – a History* Catrine, Stenlake
Savage, C.I. ((1957) *Inland Transport* London, HMSO
Scottish Divisional Council Light Railway Transport League (1951) *A Report Containing Suggestions Concerning the Future Development of Transport Services Within the City of Glasgow and its Environs*
Sharp, T. (1940) *Town Planning* Harmondsworth London, Penguin
Simmons, J. and Biddle, G. (1997) *The Oxford Companion to British Railway History* Oxford, Oxford University Press
Skillen, B.S. 'The Once and Future Railway' *Scottish Transport No.33 Special Transclyde Issue* (December 1979)
Smith, R. and Wannop, U. (eds.) (1985) *Strategic Planning in Action – The Impact of the Clyde Regional Plan 1946–1982* Aldershot, Gower
Smith, W.A.C. and Anderson, P. (1993) *An Illustrated History of Glasgow's Railways* Oldham, Irwell Press
Spaven, D. (2015, 2nd edition) *Waverley Route – The Battle for the Borders Railway* Edinburgh, Argyll Publishing
Stone, R (2002). 'Counting the Cost of London's Killer Smog' *Science* 298 (5601): pp.2106–2107
Strathclyde Partnership for Transport *Clyde Metro* (2026) https://www.spt.co.uk/about-us/what-we-are-doing/clyde-metro/
Summerfield, P. (1981) 'Education and Politics in the British Armed Forces in the Second World War' *International Review of Social History* Vol.26 No.2
Thomas, J. (1964) *The Springburn Story* Newton Abbot, David & Charles
Thomas, J. (1971) *A Regional History of the Railways of Great Britain Vol.6 Scotland – the Lowlands and Borders* Newton Abbot, David & Charles
Thorpe, A. (1995) 'Britain' in Noakes, J. (ed.) *The Civilian in War – The Home Front in Europe, Japan and the USA in World War II* Exeter, University of Exeter Press
Transport for Greater Manchester *Bee Network* (2026) https://tfgm.com/
Turbett, C. (2021) *The Anglo-Soviet Alliance – Comrades and Allies During* WW2 Barnsley, Pen & Sword
Walker, G. (1947 2nd edition) *Road and Rail – an inquiry into the economics of competition and state control* London, George Allen and Unwin
Welch, D. (2016) *Persuading the People – British Propaganda in World War II* London, British Library
William, W.E. (July 1943) *When the Lights Go On ABCA Pamphlet No.43* London, War Office
Wolmar, C. (2001) *Broken Rails – How Privatisation Wrecked Britain's Railways* London, Aurum Press
Wolmar, C. (2010) *Engines of War- How Wars Were Won & Lost on the Railways* London, Atlantic Books
Wolmar, C. (2022) *British Rail – The Making and Breaking of Our Trains* London, Penguin

NOTES

Introduction

1. Beveridge, W. (1942) *Social Insurance and Allied Services Cmd 6404* London HMSO
2. Johnston, T. (1943) *Report of the Committee on Hydro-Electric Development in Scotland Cmd 6406* Edinburgh HMSO
3. Fowler, J., Edwards, R. and Wilson, J. (2025) 'Britain's First Net Zero: Turning the Lights On and Railways Off 1953–73' *Enterprise & Society* (2025): pp.1-29
4. Dennis, G. (2024) *How the Railways Will Fix the Future* London, Repeater

Chapter 1

1. McKinnon, J. (1921) *The Social and Industrial History of Scotland: From the Union to the Present Time* London, Longman; Smith, W.A.C. & Anderson, P. (1993) *An Illustrated History of Glasgow's Railways* Oldham, Irwell Press. This last book offers the best short history of Glasgow's railways available.
2. Paterson, A.J. (1969) *The Golden Years of the Clyde Steamers (1889–1914)* Newton Abbot, David & Charles
3. Malm, A. (2016) *Fossil Capital – The Rise of Steam Power and the Roots of Global Warming* London, Verso
4. A notable exception is Ross, D. (2018) *Scottish Railways 1923–2016 – a History* Catrine, Stenlake
5. Macleod, A.B. (1948) *The McIntosh Locomotives of the Caledonian Railway* London, Ian Allan
6. Johnston, T. (1920) *The History of the Working Classes in Scotland* Glasgow, Forward Publishing pp.358–359
7. *The Railway Year Book 1922* London, Railway Publishing Company
8. Quoted in Johnston, T. (1920), as 6 above, p.360
9. Thomas, J. (1964) *The Springburn Story* Newton Abbot, David & Charles
10. Nock, O.S. (1973) *The Caledonian Railway* London, Ian Allan p.7
11. Little, S.M. 'Greater Glasgow's Railway Network' *Scottish Transport No.33 Special Transclyde Issue* (December 1979)
12. Smith & Anderson 1993, as 1 above
13. Skillen, B.S. 'The Once and Future Railway' *Scottish Transport No.33 Special Transclyde Issue* (December 1979)
14. Quoted in Smith, R. and Wannop, U. (eds.) (1985) *Strategic Planning in Action – The Impact of the Clyde Regional Plan 1946–1982* p.147, Aldershot, Gower
15. North British Railway (1894) *Mountain Moor and Loch* London, Sir Joseph Causton & Sons
16. Little (1979), as 11 above
17. Thomas, J. (1971) *A Regional History of the Railways of Great Britain Vol.6 Scotland – the Lowlands and Borders* Newton Abbot, David & Charles
18. Ross, D. (2018) *Scottish Railways 1923–2016 – a History* Catrine, Stenlake
19. Darwin, B. (1946) *War on the Line – The Southern Railway in Wartime* London, Southern Railway
20. Ministry of Transport (1931) *Report of the Committee on Main Line Electrification* London, HMSO
21. Kernahan, J. (2011 2nd ed.) *The Cathcart Circle* Lydney, Lightmoor Press
22. Ross, D. (2018) as 18 above
23. Ross, D (2018) as 18 above
24. Hutt, A. (1937) *The Post-War History of the British Working Class* London, Gollancz
25. Quoted in Johnston (1911) p285, as 6 above
26. Meighan, M. (2024) *Glasgow – A Transport History* Stroud, Amberley
27. Deans, B.T. 'Glasgow Subway in the 1920s' *Scottish Transport No.33 Special Transclyde Issue* (December 1979)
28. Savage, C.I. (1957) *Inland Transport* London, HMSO p103 (Kindle edition) – this book represents the official history of the railways and other transport systems in Britain during the war years
29. Savage, C.I. (1957) as 28 above
30. Ross, D. (2018) as 18 above
31. *Mass Observation* was established in 1937 as an independent organisation whose aim was to survey national opinion through questionnaires, recorded conversations and the observations of volunteer recorders placed throughout Britain. During the Second World War, its finding helped shape government policy and information output.

32. Calder, A. (1969) *The People's War – Britain 1939–1945* London, Jonathan Cape
33. Nock, O.S. (1974) *Electric – Euston to Glasgow* London, Ian Allan

Chapter 2

1. *Millions Like Us – Report of the Conference of Engineering and Shipbuilding Workers called by the Shop Stewards National Council at the Stoll Theatre London, March 12 1944*, p.7 London
2. Wolmar, C. (2010) *Engines of War – How Wars Were Won & Lost on the Railways* London, Atlantic Books
3. Churchill's speech in the House of Commons 4 June 1940
4. Welch, D. (2016) *Persuading the People – British Propaganda in World War II* London, British Library
5. Quoted in Welch (2016) p.16, as 4 above
6. Owen, J. (1943) *Spotlight on the Clyde* Glasgow, Communist Party Scottish Committee
7. Communist Party (1943) *How to End the Muddle on the Railways* London; Communist Party (1944) *Transport for the People: Proposals for Post-War Organisation of British Transport Services* London, Communist Party, pp.16–17
8. Bagwell, W.S. (1963) *The Railwaymen – The History of the National Union of Railwaymen* London, George Allen & Unwin; Gordon. A. (2012) *Unity is Strength – The National Union of Railwaymen – 100 Years of Industrial Unionism* London, RMT
9. The Labour Party *Post-War Organisation of British Transport – Report by the National Executive Committee of the Labour Party* London
10. Walker, G. (1947 2nd edition) *Road and Rail – an inquiry into the economics of competition and state control* London, George Allen and Unwin
11. Bagwell, W.S. (1963) as 8 above, quoted on p.595
12. *Facts About British Railways in Wartime* (1943), *British Railways in Peace and War* (1944), *It Can Now Be Revealed* (1945), *The LMS at War* (1946), *War on the Line – the Southern Railway in Wartime* (1946), *By Rail to Victory – the Story of the LNER in Wartime* (1947)
13. Hilton, W.S. (1986) *The Plug Dropper* London, Trade Press Publications p.28
14. Thorpe, A. (1995) 'Britain' in Noakes, J. (ed.) *The Civilian in War – The Home Front in Europe, Japan and the USA in World War II* Exeter, University of Exeter Press
15. Summerfield, P. (1981) 'Education and Politics in the British Armed Forces in the Second World War' *International Review of Social History* Vol.26. No.2.
16. Army Bureau of Current Affairs (November 1943) *Basic ABCA Bulletin* London, War Office
17. Kisch, R. (1985) *The Days of the Good Soldiers – Communists in the Armed Forces WWII* London, Journeyman
18. Turbett, C. (2021) *The Anglo-Soviet Alliance – Comrades and Allies During WW2* Barnsley, Pen & Sword
19. William, W.E. (July 1943) *When the Lights Go On ABCA Pamphlet No.43* London, War Office
20. Felton, Dr M. (February 1946) *It's Everybody's Business – Local Government in England ABCA Pamphlet No.115* London, War Office
21. Quoted in Calder, A. (1969) *The People's War – Britain 1939–1945* London, Jonathan Cape p.251
22. Baker, A. (1989) *The Cairo Parliament 1943–44 – An experiment in military democracy* Leigh on Sea, Partizan Press
23. Calder, A. (1969) as 21 above, p.251
24. Mackenzie, S.P. (1992) *Politics and Military Morale – Current-Affairs and Citizenship Education in the British Army 1914–1950* Oxford, Clarendon Press, and Calder, A. (1969) as 21 above
25. Sharp, T. (1940) *Town Planning* Harmondsworth, Penguin
26. Army Bureau of Current Affairs (no author cited) (September 1942) *Town Planning ABCA No.27* London, War Office
27. Calder, A. (1969) as 21 above, p.296
28. *Report of the Committee on Hydro-Electric Development in Scotland* (December 1942) Cmd 6406, Edinburgh, HMSO
29. Calder, A. (1969) as 21 above, p.584
30. *Corporation of the City of Glasgow First Planning Report* (1945) (known as the 'Bruce Report'. A copy can be viewed in the Mitchell Library, Glasgow.
31. Abercrombie, Sir P. and Matthew, R.H. (1946) *The Clyde Valley Regional Plan* Glasgow, University Press
32. Florian, U. (2010) *Modernising Glasgow – Tower Blocks, Motorways and New Towns 1940–2010* Glasgow School of Art online research repository https://radar.gsa.ac.uk/5782/
33. William, W.E. (July 1943) *When the Lights Go On ABCA Pamphlet No.43* as 19 above, p.16

Chapter 3

1. *Corporation of the City of Glasgow First Planning Report* (1945) (AKA 'Bruce Report')
2. Mitchell, A. (1995) *Election '45 – Reflections on the Revolution in Britain* London, Bellow Publishing, p.71–72
3. Kynaston, D. (2007) *Austerity Britain 1945–51* London, Bloomsbury
4. Mackenzie, S.P. (1992) *Politics and Military Morale – Current Affairs and Citizenship Education in the British Army 1914–1950* Oxford, Clarendon Press
5. 1945 Labour Manifesto: *Let Us Face the Future – A Declaration of Labour Policy for the Consideration of the Nation* http://www.labour-party.org.uk/manifestos/1945/1945-labour-manifesto.shtml (accessed November 2025)
6. Lamont, A. (1945) *How Scotland Lost Her Railways – and the case for Nationalisation on a Scottish basis* Glasgow, The Scottish Secretariat
7. Figures quoted in Wolmar, C. (2022) *British Rail – The Making and Breaking of Our Trains* London, Penguin
8. Bagwell, W.S. (1963) *The Railwaymen – The History of the National Union of Railwaymen* London, George Allen & Unwin, p.603
9. Bagwell, W.S. (1963) as 8 above
10. Kynaston, D. (2007) as 3 above, p.428
11. Wolmar, C. (2022) as 7 above, and Faulkner, R., Austin, C. (2023) *Signals Passed at Danger – Railway Power and Politics in Britain* Manchester, Crecy Publishing
12. Kynaston, D. (2007 as 3 above, p.102
13. Kynaston, D. (2007) as 3 above, p.402
14. Faulkner, R., Austin, C. (2023) as 11 above
15. British Transport Commission *Report and Accounts for 1948* London, HMSO, p.75
16. *Electrification of Railways – Report of a Committee Appointed by the Railway Executive and the London Transport Executive* (1951) London, British Transport Commission
17. Fitzpayne, E.R.L (1948*) A Report on the Future Development of Passenger Transport in Glasgow.* Copy available in the National Library of Scotland
18. Fitzpayne, E.R.L. (1948) as 17 above, pp.12–16
19. Abrams, L., Kearns, A. Hazley, B. & Wright, V. (2020) *Glasgow – High Rise Homes, Estates and Communities in the Post-War Period* London, Routledge

Chapter 4

1. Glasgow and District Transport Committee (1951) Passenger Transport in Glasgow and District London, British Transport Commission (Inglis Report), p.9
2. Festival of Britain Guide (1951) p.6
3. Your British Railways 1951 British Railways, p.1
4. Stone, R (2002) 'Counting the Cost of London's Killer Smog' Science. 298 (5601): pp.2106–2107
5. The impact of replacing coal with electricity by extension of the National Grid, and its impact on railways is argued in Fowler, J., Edwards, R. and Wilson, J. (2025) 'Britain's First Net Zero: Turning the Lights On and Railways Off 1953-73' Enterprise & Society (2025): pp.1-29
6. Inglis Report (1951), as 1 above
7. Inglis Report (1951), as 1 above, p.iii
8. Royal Commission on Transport (1930) Final Report – The Co-ordination and Development of Transport London, HMSO, p.105
9. Scottish Divisional Council Light Railway Transport League (1951) A Report Containing Suggestions Concerning the Future Development of Transport Services Within the City of Glasgow and Its Environs
10. Inglis Report (1951) as 1 above, pp.79–81
11. Railway Magazine June 1958, and Murray-Smith, D. Aberdeen to Ballater by BMU – Notes on the Battery Railcar Experiment https://www.railscot.co.uk/articles/Aberdeen_to_Ballater_by_BMU:_Notes_on_the_Battery_Railcar_Experiment/ (accessed December 2025)
12. Somewhat gleefully reported in Commercial Motor 1 February 1952
13. Mullay, A. (2006) Scottish Region, A History 1948–1973 Stroud, Tempus
14. British Railways Scottish Region (1954) Report of the Regional Committee on the Electrification of Glasgow Suburban Lines (a copy of this report is in the National Rail Museum Archive, York – part of the Michael Bonavia Collection)
15. Clough, D. (2014) The Modernisation Plan – British Railways' Blueprint for the Future London, Ian Allan – is an example of criticism of nationalisation
16. British Transport Commission (1955) Modernisation and Re-Equipment of British Railways London, British Transport Commission (the Modernisation Plan)

17. Modernisation Plan (1955) p.5, as 16 above
18. British Transport Commission (1956) Modernisation of British Railways – The System of Electrification for British Railways London, British Transport Commission
19. For example: Patten, B. (1994) Glasgow in the Age of the Tram, 1950-62 Kettering, Silver Link
20. Report of the General Manager Year 1st June 1955 to 31st May 1956 (1956) Glasgow Corporation Transport Committee
21. Oakley, C.A. The Last Tram (1962) Corporation of the City of Glasgow, p.124
22. Highway Statistics (1965) Ministry of Transport, London
23. Hansard (3rd April 1950) https://api.parliament.uk/historic-hansard/commons/1950/apr/03/consultative-committees (accessed December 2025)
24. Hansard (18th June 1958) https://api.parliament.uk/historic-hansard/commons/1958/jun/18/transport-users-consultative-committees (accessed December 2025)
25. Florian, U. (2010) Modernising Glasgow – Tower Blocks, Motorways and New Towns 1940–2010 Glasgow School of Art online research repository https://radar.gsa.ac.uk/5782/ (accessed November 2025)
26. Robertson, D.S. (1998) 'Pulling in opposite directions: the failure of post-war planning to regenerate Glasgow' Planning Perspectives 13.1 pp53–67

Chapter 5

1. Blake, G. (1960) Glasgow Electric – the story of Scotland's electric railway British Railways Scotland
2. Bond, R.C. (1975) A Lifetime with Locomotives Cambridge, Goose & Son
3. Scotsman 15 February and 16 March 1956
4. Railway Magazine June 1956
5. Railway Magazine January/February 1947, quoted in Haresnape, B. (1983) British Railways 1948-83 – A Journey by Design London, Ian Allan, p.47
6. Cousins, J. (1986) British Rail Design Copenhagen, Danish Design Council – a book that exemplifies this recognition
7. Clough, D.N. (2014) The Modernisation Plan – British Railways Blueprint for the Future London, Ian Allan, p.110
8. Ross, D. (2018) Scottish Railways 1923-2016 – A History Catrine, Stenlake, p.120
9. Bond, R.C. (1975) as 2 above
10. Gourvish, T.R. (1986) British Railways 1948-73 – A Business History Cambridge, Cambridge University Press
11. 'Glasgow's Rail Lead – Suburban Electric Trains to be best in world' Glasgow Herald 4 September 1959
12. Ross, D. (2018) Scottish Railways 1923–2016 – A History Catrine, Stenlake, p.86
13. Blake, G. (1960) Glasgow Electric – the story of Scotland's electric railway, as 1 above
14. 'Switch is on to the Blue Trains.' Evening Times 7 November 1960
15. BR Public Relations Division (undated but probably 1962) Internal Relations Case Study – The Glasgow Blue Trains British Railways, p.15
16. BR Public Relations Division as 15 above
17. Much of the detail in this section concerning the 1960 failures and subsequent remedial work comes from the official Ministry of Transport report Failures of Electric Multiple Units on British Railways Final Report (1962) London, HMSO, where a very full technical account is given.
18. Beharrell, L.V. (1962) 'Glasgow Blue Trains: Some Implications' British Transport Review VI (5) April 1962, p.363
19. Campbell, R.W. (1919) Snooker Tam of the Cathcart Railway Edinburgh, W.&R. Chambers p.2
20. Kernahan, J. (1980) The Cathcart Circle Falkirk, Scottish Railway Preservation Society
21. Millar, A. (1985) British PTEs: 1 – Strathclyde London, Ian Allan
22. Official accident reports from which this section is drawn can be found on the Rail Archive website: https://www.railwaysarchive.co.uk/
23. Bonavia, M.R. (1981) British Rail – The First 25 Years Newton Abbot, David & Charles
24. British Railways Board (1966) Annual Report and Accounts 1965 London, HMSO, p.6

Chapter 6

1. British Railways Board (1973) British Railways Board: Report and Accounts 1972
2. Bonavia, M.R. (1981) British Rail – The First 25 Years Newton Abbot, David & Charles, pp.110–113
3. British Railways Board (1963) The Reshaping of Britain's Railways (known as the 'Beeching Report') London, HMSO
4. National Union of Railwaymen (1963) The Misshaping of Britain's Railways NUR, London
5. Booth, J. (2025) Rail Workers 200: Two Centuries of Railworkers and their Struggles London, Workers Liberty Publications

6. These figures, from official sources, are quoted from Bonavia, M.R. (1981) British Rail – The First 25 Years (as 2. above), and Bagwell, P.S. (1982) The Railwaymen – The History of the National Union of Railwaymen Vol.2: The Beeching Years and After London, George Allen & Unwin
7. Austin, C., Faulkner, R. (2015) Disconnected – Broken Links in Britain's Rail Policy Surrey, OPC
8. Bonavia, M.R. (1981) British Rail – The First 25 Years, as 2 above, p.128
9. Gourvish, T.R. (1986) British Railways 1948-73 – A Business History Cambridge, Cambridge University Press, p.382
10. British Railways Board (1973) Report and Accounts 1972
11. Wolmar, C. (2022) British Rail – The Making and Breaking of Our Trains London, Penguin
12. Buxton, J., Heath, D. (2025) Lines of Power – The Electrification of British Railways – Trials Tribulations and Triumphs Lewes, Unicorn
13. Daniel, A.W.T., Calvert, R. (November 1969) Electrify Now – A compelling case for further railway electrification in Great Britain, especially between Crewe and Glasgow London, National Council for Inland Transport
14. British Rail (1978) Railway Electrification: A British Railways Board Discussion Paper
15. NUR (1979) Railway Electrification – Submission to the Secretary of State for Transport from the National Union of Railwaymen London, NUR
16. Buckton, R. (1982) Transport – The Forgotten Social Service – lecture given to the Chartered Institute of Public Transport by Ray Buckton, General Secretary, ASLEF
17. Wolmar, C. (2022) British Rail – The Making and Breaking of Our Trains, as 11 above
18. Faulkner, R., Austin, C. (2023) Signals Passed at Danger – Railway Power and Politics in Britain Manchester, Crecy
19. For example: Ross, D. (2018) Scottish Railways 1923-2016 – A History Catrine, Stenlake; Wolmar (as 11 and 17 above); Faulkner, R., Austin, C. (2023) Signals Passed at Danger – Railway Power and Politics in Britain, as 18 above
20. Wolmar, C. (2001) Broken Rails – How Privatisation Wrecked Britain's Railways London, Aurum Press

Chapter 7

1. Foreword to: Noble, T. (1990) Exploring Strathclyde by Rail North Yorkshire, Leading Edge
2. Corporation of the City of Glasgow, Scottish Development Department (1966) Greater Glasgow Transportation Study Vol.1: Current Travel Patterns in the City of Glasgow – Origin Destination Survey 1964–65; Corporation of the City of Glasgow, Scottish Development Department (1968) Greater Glasgow Transportation Study Vol.2: Forecast and Plan. I have also borrowed analysis of the GGTS work from Millar, A. (1986) British SPTs 1: Strathclyde London, Ian Allan
3. Millar, A. (1986) British SPTs 1: Strathclyde, as 2 above p.26
4. Walker, P.J. (1974) 'Light rapid transport for Glasgow' Modern Tramway and Rapid Transit Review June 1974
5. Greater Glasgow Passenger Transport Executive (1974) Development of CLYDERAIL – Application to Secretary of State for Scotland for Infrastructure Grant Glasgow, GGPTE
6. Craig, G.B., Logie, J.G., Scott, K.D. and Taylor, M.S. (1980) 'The Transclyde railway network: Kelvinhaugh burrowing junction' Proceedings of the Institution of Civil Engineers Part 1 1980, August, pp.369–389
7. Greater Glasgow Passenger Transport Executive (1979) Trans-Clyde News No.3
8. British Rail, Strathclyde Transport (1983) Review of Strathclyde Rail Network – the Way Ahead
9. Ross, D. (2018) Scottish Railways 1923-2016 – A History Catrine, Stenlake
10. Berlin, M. (2006) Never on Our Knees – A History of the RMT, 1979–2006 London, Pluto
11. Ross, D. (2018) Scottish Railways 1923-2016 – A History, p.154; Stevenson, G. (2000) British Bus Deregulation London, International Transport Workers' Federation
12. Report by Working Group on Supported Rail Services (September 1987) Glasgow, Strathclyde Regional Council
13. Public Transport in Strathclyde Region – The Way Ahead – Issues and Options for Assessment (April 1989) Glasgow, Strathclyde Regional Council
14. Public Transport for the 21st Century – Strathclyde Transport Development Study Consultation Document (December 1989) Glasgow, Strathclyde Regional Council
15. Abbis, J. and Lumsdon, L. (1979) Route Causes – A Guide to Public Transport Plans London, Transport 2000/ National Council of Social Service
16. Wilson, J.S. (1990) 'Strathclyde Studies Light Rail' Modern Tramway September 1990
17. 'Tram Plan Hits the Rails' (1996) Scotland on Sunday newspaper, 25 August 1996

Chapter 8

1. Scotland's Climate Change Assembly – Recommendations for Action (June 2021) Scotland's Climate Assembly Recommendations for Action.WebVersion (2) (2).pdf

2. Understanding Glasgow (2026) https://www.understandingglasgow.com/glasgow-indicators/transport/vehicle-availability/scottish-cities
3. Curl, A., Julie Clark, J. and Kearns, A. (2018) ' Household car adoption and financial distress in deprived urban communities: A case of forced car ownership?' Transport Policy 65, pp.61–71
4. Figures from Ross, D. (2018) Scottish Railways 1923–2016 – A History Catrine, Stenlake, p.199; Transport Scotland Scottish Transport Statistics 2024 https://www.transport.gov.scot/media/r4dn22ko/chapter-1-road-transport-vehicles-scottish-transport-statistics-2024.pdf ; ORR Key Train Statistics 2024–2025 https://dataportal.orr.gov.uk/media/fjdjeejh/scotrail-key-stats.pdf
5. Ross, D. (2018) Scottish Railways 1923–2016 – A History, as 4 above
6. Buxton, J. and Heath, D. (2025) Lines of Power – The Electrification of British Railways – Trials, tribulations and triumphs Lewis, Unicorn
7. Glen, A. (2010) Airdrie–Bathgate Rail Link – Reconnecting Communities Isle of Man, Lily
8. Transport Scotland (2013) Airdrie–Bathgate Rail Link Improvement – Stage 1 Outcome Evaluation Report https://www.transport.gov.scot/media/32477/j374405.pdf
9. Glen, A. (2018) Transforming the Railways of Central Scotland Isle of Man, Lily; Buxton, J. and Heath, D, (2025) Lines of Power, as 6 above
10. Glen, A. (2020) Glasgow Queen Street – A Railway Station Renaissance Isle of Man, Lily
11. Spaven, D. (2015, 2nd edition) Waverley Route – The Battle for the Borders Railway Edinburgh, Argyll Publishing
12. Scottish Government (2021): https://www.gov.scot/publications/national-mission-local-impact-infrastructure-investment-plan-scotland-2021-22-2025-26/; Scottish Government (2024) https://www.transport.gov.scot/publication/scottish-transport-statistics-2024/chapter-10-finance/
13. Scottish Government (2025) Rail Recharged – Scotland's Fleet Replacement Strategy https://www.transport.gov.scot/media/5z2k050m/rail-recharged-scotlands-fleet-transition-strategy-pdf-version.pdf
14. Transport for Greater Manchester Bee Network (2026) https://tfgm.com/
15. I am grateful to Get Glasgow Moving for the analysis used in this chapter https://www.getglasgowmoving.org/
16. Strathclyde Partnership for Transport (2017) Options to Extend Glasgow Subway https://www.getglasgowmoving.org/reports/extendglasgowsubway.pdf
17. Strathclyde Partnership for Transport Clyde Metro (2026) https://www.spt.co.uk/about-us/what-we-are-doing/clyde-metro/
18. 'Hydrogen's Future in Rail's Hands.' Rail 7 January 2026

INDEX